D1329651

Coal Conversion Technology

Coal Conversion Technology

Problems and Solutions

Arnold H. Pelofsky, EDITOR
Science Applications, Inc.

Based on a symposium sponsored by the ACS Division of Industrial and Engineering Chemistry at the I&EC Winter Symposium, Colorado Springs, Colorado, February 12–13, 1979

ACS SYMPOSIUM SERIES **110**

AMERICAN CHEMICAL SOCIETY
WASHINGTON, D. C. 1979

Library of Congress CIP Data

Coal conversion technology.
(ACS symposium series; 110 ISSN 0097-6156)

Includes bibliographies and index.

1. Coal liquefaction—Congresses. 2. Coal gasification—Congresses.
I. Pelofsky, Arnold H. II. American Chemical Society. Division of Industrial and Engineering Chemistry. III. Series: American Chemical Society. ACS symposium series; 110.

TP352.C63 662'.625 79-17936
ISBN 0-8412-0516-7 ACSMC8 110 1-257 1979

PRINTED IN THE UNITED STATES OF AMERICA

DEDICATION

This book is dedicated to all of the individuals who have spent all or most of their working lives trying to persuade the "powers-that-be" that the conversion of coal to other useful forms of energy is the thing to do.

ACS Symposium Series

M. Joan Comstock, *Series Editor*

FOREWORD

The ACS Symposium Series was founded in 1974 to provide a medium for publishing symposia quickly in book form. The format of the Series parallels that of the continuing Advances in Chemistry Series except that in order to save time the papers are not typeset but are reproduced as they are submitted by the authors in camera-ready form. Papers are selected to maintain the integrity of the symposia; however, verbatim reproductions of previously published papers are not accepted. Both reviews and reports of research are acceptable since symposia may embrace both types of presentation.

CONTENTS

PREFACE

The United States has more Btu's in its coal reserves than the Mid-East has in its oil reserves. The United States, if it is to approach self-sufficiency, must exploit coal. Since there are transportation systems for oil and gas products already available in the United States, it would make obvious sense to convert coal into these products so that these usable forms of energy could be shipped to the end-user. Unfortunately, there are technical, institutional, and financial barriers that have prevented the development of the coal conversion industry. These problems and potential solutions to them are delineated in this book.

This book includes a compendium of papers presented at a symposium entitled Coal Dilemma II and the discussions that followed between the authors and the participants. The objective of the symposium was to present problems and postulate solutions. The papers are the most current in coal conversion. Technical and economical information is presented in all the papers that appear. Probably the most important aspect of this book is the discussions that followed the presentations of the papers by the participants in the audience and the authors. These discussions will give the reader an insight to the complex nature of the problems that are faced by the United States.

I would like to acknowledge the services of and thank several individuals without whose help this manuscript could not have been prepared. Arthur Conn, President of Arthur L. Conn & Associates, and Leonard Seglin, President of Econergy Associates, were the cochairmen of the two-day symposium. They enlisted the aid of the authors whose manuscripts are included in this text and helped stimulate interest in the technical community. I would also like to acknowledge Rosemary Szymanski, Suzanne Rigler, and Loretta Pelofsky for typing, proofreading, and generally preparing the manuscript for publication. Last, but not least, I would like to thank the Division of Industrial and Engineering Chemistry for honoring me by giving me the opportunity to be the general chairman of the symposium and the editor of this manuscript.

Science Applications, Inc.
East Brunswick, New Jersey
May 21, 1979

ARNOLD H. PELOFSKY

CONTRIBUTORS

RALPH BLOOM, JR., Manager of Business Development, COGAS Development Company

GENERAL JAY R. BRILL, Deputy Undersecretary for Strategic Petroleum Reserves, Department of Energy

E. L. CLARK, Consultant

H. D. COCHRAN, Coal Conversion, Oak Ridge National Laboratory

W. ROBERT EPPERLY, General Manager, Exxon Research & Engineering Company

GERARD C. GAMBS, Vice President, Ford, Bacon & Davis, Inc.

RICHARD F. HILL, Executive Manager, Engineering Societies Commission on Energy

DR. HENRY R. LINDEN, Professor of Gas Engineering and Research, Professor of Chemical Engineering, Illinois Institute of Technology

RICHARD A. PASSMAN, Director, Office of Coal Resource Management U.S. Dept. of Energy

BRUCE K. SCHMID, Technical Advisor, Gulf Mineral Resources

HOWARD M. SIEGEL, Manager, Synthetic Fuels Engineering Dept., Exxon Research & Engineering Company

J. C. SWAN, Project Director, Ashland Synthetic Fuels, Ashland Oil Company

RONALD H. WOLK, Program Manager, Electric Power Research Institute

CONTRIBUTORS

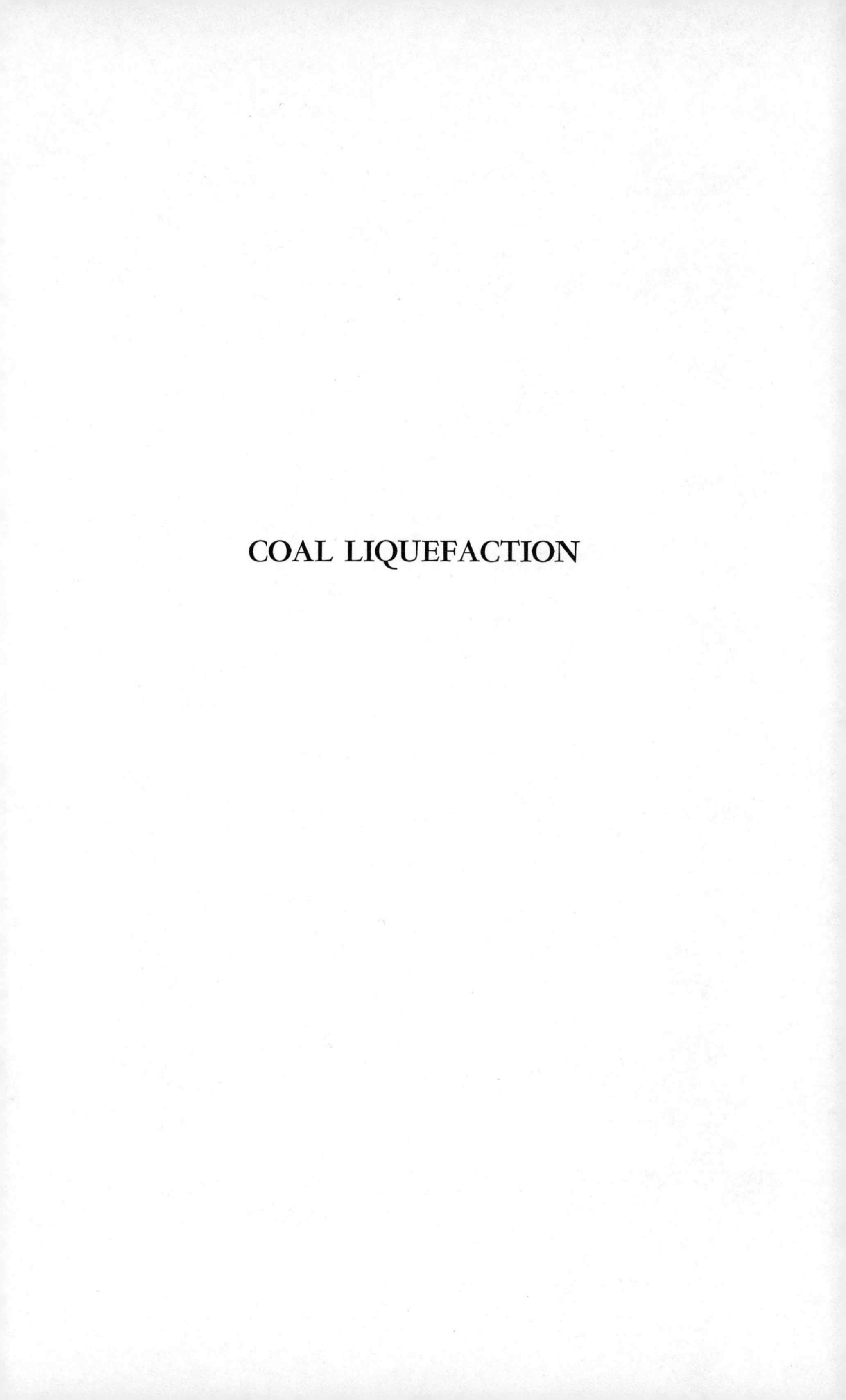

COAL LIQUEFACTION

Section Introduction

CHAIRMAN SEGLIN: As you will note in the printed program, the program is divided into three parts. One is to hear from one of the major customers of liquids from coal; this is the utility industry, and the second part is to get a status report from each of the major approaches to coal liquefaction, with the third part being a discussion period. The discussion phase will include questions and answers after each status report, if time permits, and more intensive discussion this afternoon, when the panel of speakers will meet on the stage.

The last part is an instant overview of the subject by myself. This is to set the stage for the Coal Dilemma vis-a-vis liquefaction.

The major question as I see it, and I'm sure that there will be differences of opinion, is: Can the high cost of synthetic fuel be justified and when? My judgement, which is based on considerable thought, is "yes". There is enormous pressure to increase the price of oil, and as long as you have a monopoly situation, such as OPEC, I think that will prevail. The estimated price-driving force, based on current market experience, shows that the price estimates for synthetic "crude" is less than the unrestrained OPEC oil price potential. Hence, it appears that at some time in the future, the price of crude and the price of syncrude will cross, and synthetic fuels will be acceptable in the marketplace.

What are our options? Well, we can replace gas and oil by coal directly in large industrial furnaces and in utility boilers, or at some point we can replace gas and oil by synthetic fuels from coal or other fossil-based materials.

There are a great number of restraints in these options. What are they? Of course, the attitude of the Federal Government is a tremendous restraint. Public opinion vis-a-vis the environment presents another restraint. Thirdly, money; the financing of these giga-buck projects is another restraint.

0-8412-0516-7/79/47-110-003$05.00/0

Additionally, there are risks in the marketplace, and technical risks. The marketplace probably presents the larger uncertainty. To add to the problem, we have the cost and uncertainties due to long lead times that are present with these complex projects. Ten years is probably realistic. I don't know anybody in this audience who can state with surety what they expect to happen in the next ten years. The picture then could be significantly different from our expectations. You might say, "Well, maybe we will find another Persian Gulf and all our problems will be solved," or "Maybe we won't and we'll have to go to synthetic fuels. At some point, we will." But when you have a ten-year lead time ahead of you, it's going to be awfully hard to be the prophet in order to make any decision. I think the prophet resides in Washington--the site of most of our uncertainties.

In technology, we have many options as to what we can do and how we do it. Maybe if we had only one option, it would make our life simpler. But I think that in a way the technical options are resolvable. There are a fair number of technologies available. We will discuss some of them this morning.

What questions need answering before such a decision is made? I think one of the basic questions is make versus import. It is equivalent to the same question we have in private industry which is make versus buy. But when you look at it from the standpoint of national self-interest, I am sure there is a justifiable price premium that would justify making over and above importing. But how you assess that differential is a question with which I think the economists in their wisdom might be able to help us. Many countries arbitrarily followed the path of making rather than importing. In the long run, we might justifiably follow the same path vis-a-vis making syn-crude purely on the basis of our own self-interest. Perhaps it is about time we, as a nation, consider ourselves first.

Then there is the acceptance of the financial risk. Somebody is going to have to put one, two or three billion dollars per plant into it. Who is going to risk that kind of money, and where is it going to come from? As I said before, the marketplace is probably the rate-determining factor. I think that we, as engineers, can handle the technology. We've handled it before, and I think if we take a firm stand, we can do so in the synthetic fuel industry.

Of course, before a decision on producing synthetic fuels is made, the last question you have to answer to is that of oil politics. This is a fact of life led by OPEC, and it is reacted to by the OECD, the Organization of Economic Cooperation, of which we are a part, and by the developing nations.

1

Coal Liquefaction and the Electric Utility Industry

RONALD H. WOLK and SEYMOUR B. ALBERT

Electric Power Research Institute, 3412 Hillview Ave., Palo Alto, CA 20004

Coal liquefaction offers the utility industry an option, based on domestic energy resources, with which to meet its need for liquid fuels. In 1977, generation of electricity consumed, as shown in table 1, 188,000 BPD of distillate fuels and 1,469,000 BPD of residual oil. (1)

TABLE 1

Electric Utility Industry Use of Gaseous and Liquid Fuels (1)

	Actual 1977 000's B/D FOE	Estimated 1987 000'sB/D FOE
Distillate Oil-Steam	57	70
Combustion Turbine	116	152
Combined Cycle	15	144
Residual Oil-Steam	1,466	1,797
Combustion Turbine	1	1
Combined Cycle	2	11
Crude Oil-Steam	9	8
Sub Total	1,666	2,183
Gas - Steam	1,149	425
Combustion Turbine	23	9
Combined Cycle	37	23
Sub Total	1,209	457
Grand Total	2,875	2,640
Potential Additional Oil Needed to Compensate for 1-2 year delays in Nuclear and Coal Plant Construction		1,041

0-8412-0516-7/79/47-110-005$05.00/0

The National Electrical Reliability Council projects in their August 1978 report, that this requirement will grow to 366,000 BPD and 1,809,000 BPD respectively by 1987. In addition, natural gas requirements which can be met by the substitution of clean liquid fuels will decline from the 1977 level of 1,209,000 BPD FOE (fuel oil equivalent) to a still substantial 457,000 BPD FOE. This combination calls for 2,632,000 BPD of hydrocarbon fuels in 1987.

This same report discusses the potential for additional requirements for liquid fuels due to a one or two year delay in completion of coal and nuclear plants. If electricity growth averages 5.6% per year compounded, an additional 1,041,000 BPD could be required if such a delay occurred. The experience of 1977 where liquid fuels were utilized to cope with the combination of a severe winter that curtailed natural gas supplies used for power generation and a coal strike demonstrate that liquid fuels can be quickly utilized to meet emergency situations.

Today, the planned installation of new oil fired steam boilers is essentially nil. Table II shows that approximately 96,000 mw of capacity will remain in place in 1987. These units were put into service primarily in the mid-1960's and have 10-30 years of useful life remaining. Installed capacity of liquid fueled combined cycle units is expected to grow from 3000 to 8000 mw over this time period. These units generate electricity more efficiently than conventional boilers. Combined cycle capacity is projected to be utilized much more extensively than in the past. As a result, the anticipated quantity of power generated from combined cycle equipment may increase nine-fold from 4,000 to 36,000 million Kilowatt hours as shown in Table III. Unfortunately, the future use of petroleum liquids for this kind of operation has been jeopardized by the recently legislated Fuel Use Act. This Act requires coal to be used instead of petroleum for new power stations.

Liquid fuels are desirable to utilities because they are:

- o clean and satisfy environmental restrictions
- o readily storable and transportable
- o have properties that can be tailored to meet user requirements and
- o can be used in new combustion turbines and combined cycle machines to meet intermediate and peaking power requirements at less cost than coal fired plants.

Although the prices of petroleum derived liquid fuels are significantly higher than coal and nuclear fuel, the electric generating equipment to utilize them is less costly.

TABLE II

Installed Generating Capacity (1)
000's Megawatts

	Installed Capacity		Percent of Total	
	1977	1978	1977	1978
Nuclear	43	160	8.5	19.9
Hydro	59	68	11.7	8.5
Pumped Storage	10	18	2.0	2.2
Geothermal	1	2	0.2	0.2
Steam - Coal	198	343	39.1	42.7
Steam - Oil	90	96	17.8	12.0
Combustion Turbine - Oil	36	43	7.1	5.4
Combined Cycle - Oil	3	8	0.6	1.0
Steam - Gas	61	57	12.0	7.1
Combustion Turbine - Gas	3	3	0.6	0.4
Combined Cycle - Gas	2	2	0.4	0.2
Other	0	1	0	0.1
	506	803	100.0	99.7
Total Oil Fired	129	147	25.5	18.4
Total Gas Fired	66	62	13.0	7.7

TABLE III

Power Generated

	Billions KWHR Generated		Percent of Total	
	1977	1987	1977	1987
Nuclear	262	979	12.4	27.3
Hydro	220	237	10.4	6.6
Pumped Storage (Net)	(4)	(7)	(0.8)	(0.2)
Geothermal	3	15	0.2	0.4
Steam - Coal	982	1770	46.5	49.4
Steam - Oil	335	404	15.9	11.3
Combustion Turbine - Oil	18	24	0.9	0.7
Combined Cycle - Oil	4	36	0.2	1.0
Steam - Gas	277	115	13.1	3.2
Combustion Turbine - Gas	4	2	0.2	0.1
Combined Cycle - Gas	8	5	0.4	0.1
Other	3	6	0.1	0.2
	2,113	3,587	100.0	100.0
Total Oil Based	357	464	17.0	13.0
Total Gas Based	289	122	13.7	3.4

This combination makes them the least costly generating option for low and intermediate capacity factor power generation as shown in Table IV. (2) NERC projections indicate that the only major shifts anticipated in unit capacity factors will be an increase from 15% to 50% in liquid fueled combined cycle units and a decrease from 52% to 23% for gas fired boilers.

TABLE IV

Tradeoffs Between Investment and Fuel Cost

	Plant (2) Investment $/KW 1977	Fuel(2) Cost $/10^6 Btu 1977	Capacity Factor(1) 1977	1987
Nuclear	850	0.55	69.6	69.7
Steam - Coal	700	1.00	56.5	58.8
Steam - Oil	400	2.24	42.5	48.1
Combustion Turbine-Oil	150	2.57	5.7	5.9
Combined Cycle-Oil			15.3	49.7
Steam - Gas			51.5	23.0

Coal and nuclear facilities cannot be used in a cost effective way to provide peak power generation required by a typical weekly utility demand curve as shown in Figure 1. (3) Liquid and gaseous fuels meet this need now and will be used for this type of service over the next decade or more. Utilities have serious concerns about legislation preventing the use of domestic gas or imported oil to meet these requirements. This situation leaves the utilities between the proverbial "rock and a hard place."

One alternative candidate for meeting these needs is coal derived liquids. Technology development is now proceeding along a solid path. Two large pilot plants producing liquid fuels from coal will be in operation in 1980. Successful results from these could allow the first demonstration or pioneer plants to come on stream around 1985. Assuming technological success, capacity buildup would occur as economic and/or political circumstances dictate. The establishment of a reliable supply of liquid fuels from coal for power generation then becomes a political decision, not a technical one.

All liquefaction processes produce a wide spectrum of products. Ultimately each product from a coal conversion plant will be utilized in a manner that provides the highest economic return to the plant owner. Products boiling below about 350°F will be disposed of to the transportation and petrochemical sectors of the economy. The major product in this category, aromatic naphthas, are particularly valuable as high octane gasoline blending stock.

It is anticipated that coal derived liquids boiling above about 350°F will be disposed of to the utility market. Table V summarizes the potential utility markets for various types of coal derived fuels which include solvent refined coal, heavy boiler fuels, distillate boiler fuels, turbine fuels and methanol. Speculative locations for these markets are indicated on Figure 2.

TABLE V

Fuel Type	Process	Potential Markets
Methanol		o Peaking combustion turbine
Turbine Fuels	Hydrotreated fractions from: o H-Coal o Exxon	o Combustion turbines o Intermediate load combined cycle units
Distillate Boiler Fuels	Fractions from: o H-Coal o Exxon Donor Solvent o SRC-II	o Retrofit gas fired boilers o Retrofit oil boilers for peaking service
Heavy Liquid Boiler Fuels	Fractions From: o H-Coal o Exxon Donor Solvent	o Retrofit existing oil fired base load units
Solid Boiler Fuel	Solvent Refined Coal	o Retrofit existing intermediate load plant o Specifically designed simplified base load plants

Coastal utilities have been major consumers of products derived from imported crudes. East coast utility fuels have been based on Venezuelan and Middle East crudes while the West coast has obtained much of its fuel from Indonesia. There are a number of reasons why it would be difficult to convert these plants to coal firing. Auxiliary facilities such as storage areas, rail sidings, and unloading and conveying equipment are no longer in place to handle coal. It is even more significant that the land on which these facilities were located has been sold or used for other utility purposes. As a result, scrubbers could not be installed at these sites to allow for sulfur dioxide control.

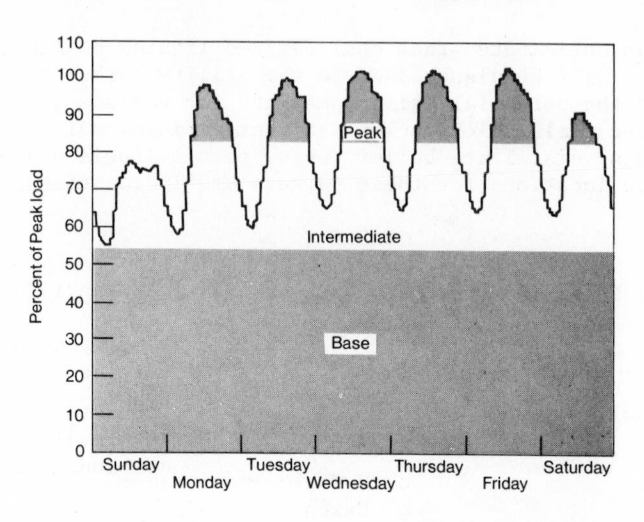

Figure 1. Weekly load curve

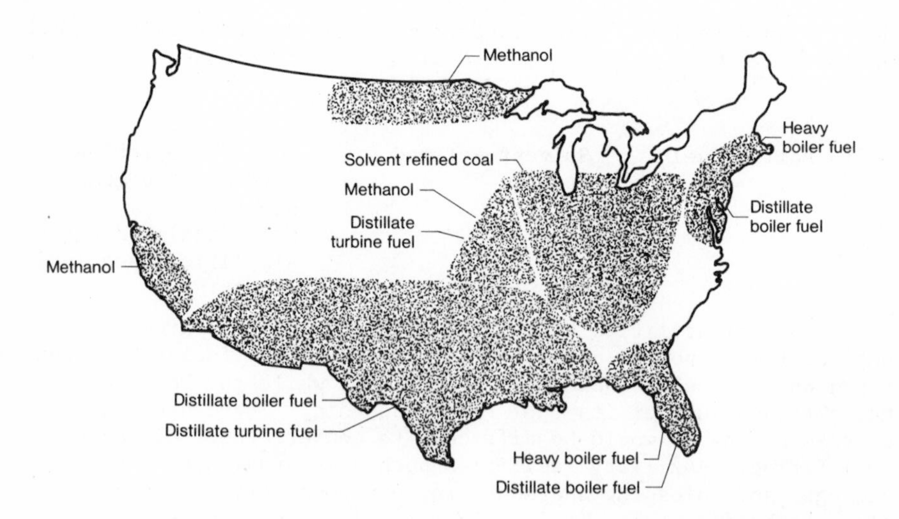

Figure 2. Potential coal liquefaction production markets

Coal is not a viable option in many urban areas because of very stringent emission standards for nitrogen oxides and particulate emissions. However, coal derived liquid feuls with tailored properties could be used to meet these requirements.

Midwest utilities are coal burners. They have the know-how and facilities to utilize solid fuels. Solvent refined coal, which has the potential of being the lowest cost coal liquefaction product because of its low hydrogen content, is of interest to this group.

Many of the Southwest states have a large number of gas fired boilers. These units were very low in original investment cost and their continued utilization concerns utilities in those areas. Hydrotreated coal derived distillates offer a means of keeping these units available for years of additional service.

Scattered areas in the country with very stringent emission standards and very sharp peaks in electricity demand may be able to justify methanol for peaking service in minimum capacity factor service.

Whether or not there is actual utilization of these products in these markets will depend on a number of factors:

- o availability of alternate fuels
- o environmental regulations
- o fuel price
- o government regulations concerning utilization

Price of raw coal derived liquid products will likely be in the range of \$3.50-\$5.00 per million Btu's in 1978 dollars. (4) Extensive hydrotreating to reduce heteroatom content may add on the order of \$1-2 per million Btu's. (5) Typical costs for this upgrading step are presented in Figure III. Economic projections indicate that these costs can reac- price parity with petroleum derived fuels sometime between 1985 and 1995.

There is a wide support in the utility industry for the development of a number of liquefaction processes. In this way the probability of technical success for the overall objective is enhanced. Another benefit which is not so apparent is the avoidance on development of a single process which may not be applicabel to a wide variety of commercially important coals.

There is no evidence that we are aware of to indicate that any single liquefaction process offers a significant economic advantage over all others if the desired product slate is fixed. At our current level of understanding, all leading process candidates, H-Coal, Exxon Donor Solvent, and SRC-II all appear to

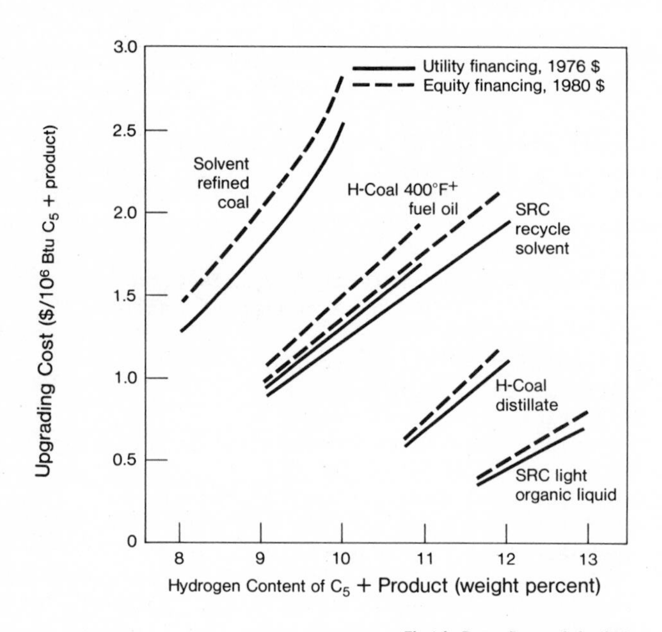

Electric Power Research Institute

Figure 3. Cost of upgrading coal liquids to turbine fuel

produce a specified slate of products at approximately the same cost from a given coal. The uncertainty in the costs based on assumptions of engineering requirements is larger than the difference between processes producing similar product slates and quality.

Combustion Testing Programs

The utility industry requires comprehensive, large scale, and long duration tests in utility equipment prior to accepting any new fuel. As an example, the changeover from eastern coal to western coal was traumatic for many utilities in that a large number of new maintenance problems and emission control difficulties were generated. In line with these requirements EPRI has set up a multitiered synthetic fuel combustion test program. All new boiler fuels are first burned in small scale furnaces of 1-5 x 10^6 Btu/hr. capacity. This is followed by tests in boilers of about 50 x 10^6 Btu/hr. capacity. Data from these small scale programs are used in developing the actual test program for a utility test. A summary of the kinds of tests, contractors and fuel firing rate is presented in Table VI. This route has been followed for the large scale SRC-I and SRC-II combustion tests carried out in 1977 and 1978 respectively. Key data from these two test programs are presented in Table VII. Both utility hosts, Southern company Services Inc. and Consolidated Edison of New York, considered the tests to be successful. Unfortunately, both test programs were of relatively short duration because of the limited amount of fuel available, 300 tons of SRC-I and 4500 barrels of SRC-II. These quantities are huge in terms of the total amount of synthetic fuels generated during the last 10 years in the United States. Further testing of synthetic fuels is considered desirable and is a justification for installing first-of-a-kind pioneer and demonstration plants.

Testing of turbine fuels is handled in an analogous manner. Three sizes of test rigs have been utilized in the EPRI combustion test program-mini, sub-scale, and single combustor cans. The relative dimension of the three systems are shown in Figure IV. Combustion test data has been collected on a large number of raw and hydrotreated product samples from the SRC-I, SRC-II, Exxon Donor Solvent, H-Coal, and other processes under development. Figure V is a plot of NO_x level versus turbine inlet temperature for these fuels. The actual levels of NO_x are related to the actual piece of equipment utilized for the test series but the relative rankings are consistent among the various types of equipment.

Methanol is the most expensive of synthetic liquids that are derived from coal. Efforts are underway to reduce its cost. Its use may be justivied in combustion turbines that have the minimum

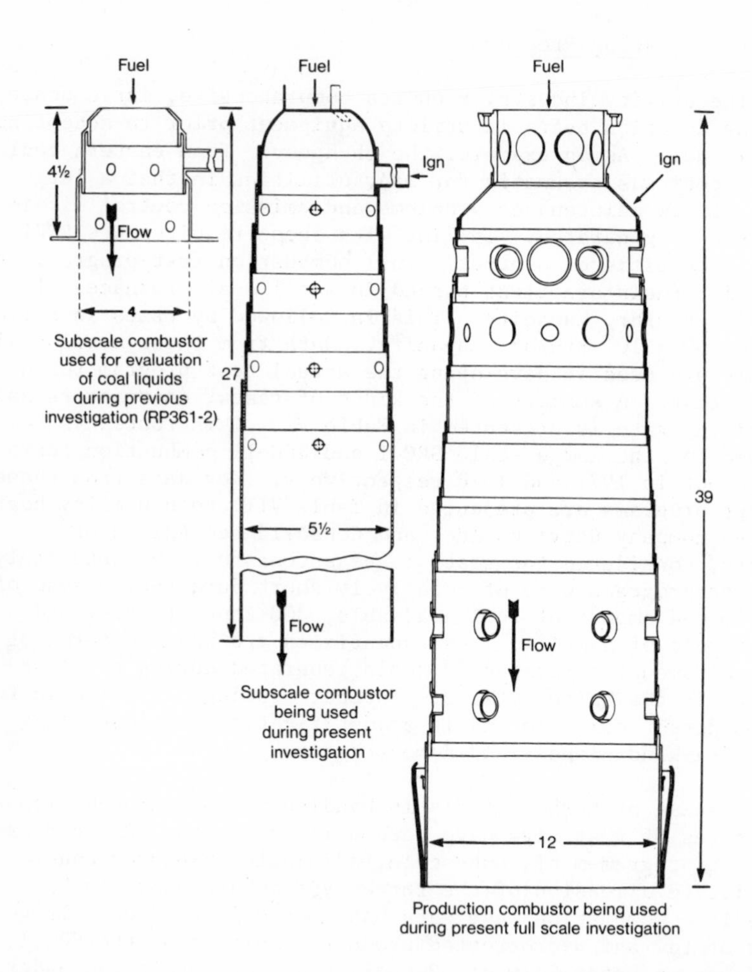

Electric Power Research Institute

Figure 4. Comparison of combustors used in evaluation of coal liquids

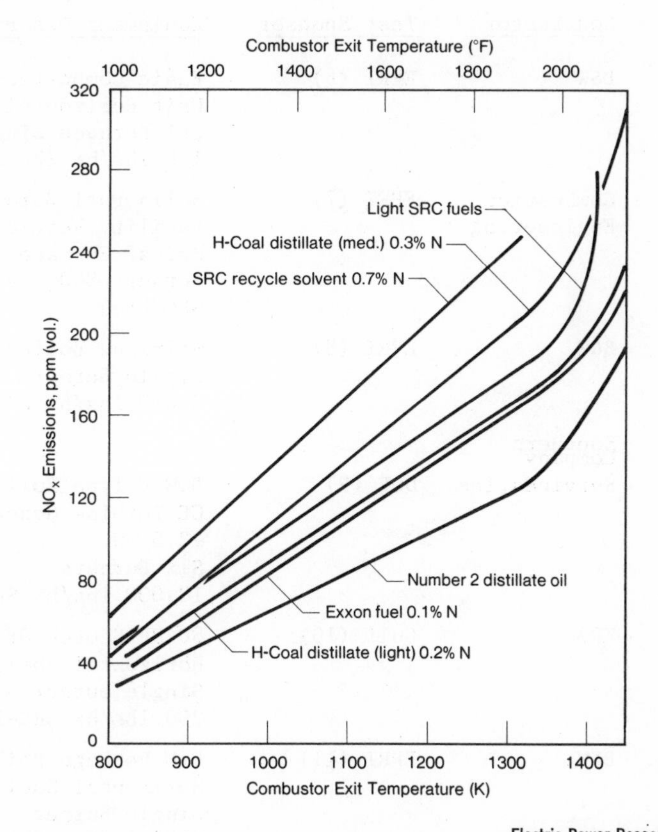

Electric Power Research Institute

Figure 5. Fuel nitrogen content as it influences subscale combustor NO_x *emissions*

load factor in a given utility system. A comparative test of methanol and Number 2 fuel oil (fuel nitrogen about 0.1%) in a gas turbine at Southern California Edison's Ellwood Station is expected to show NO_x levels 80% less than those without water injection and 20% less than those with water injection.

TABLE VI

Fuel	Contractor	Test Sponsor	Equipment Description
SRC-I	B&W	EPRI (6)	Basic Combustion Test Unit Horizontal Cylindrical Furnace Single Burner 170 lbs/hr SRC Feed
	Combustion Engineering	EPRI (7)	Solid Fuel Burning Test Facility Vertical Cylindrical Furnace Single Burner 300 lbs/hr SRC Feed
	B&W	EPRI (8)	Stirling Boiler Single Burner 3,000 lbs/hr SRC Feed
	Southern Company Services, Inc.	DOE (9)	B&W F Type Boiler GE Turbine Generator 22.5 MW Six Burners 18,000 lbs/hr SRC Feed
SRC-II	KVB	Gulf (10)	80 HP Scotch Dry Back Horizontal Shell Single Burner 200 lbs/hr SRC-II Fuel
	B&W	EPRI (11)	F M Package Boiler Horizontal Shell Single Burner 3000 lbs/hr SRC-II Fuel
	Consolidated Edison; KVB	EPRI, Con Ed. New York State ERDA (12)	Combustion Engineering 450,000 lb/hr Steam Eight Burners Two burners per corner at different elevations 25,000 lbs/hr SRC-II Fuel

Several utilities which are facing decision on how to meet peak load demands in the mid and late 1980's are looking seriously at how methanol might be used to meet those needs. Combustion turbines are relatively cheap, can be sited with less difficulty than other power generating equipment and have essentially instantaneous on-off capability. Utilization of a super clean fuel such as methanol may be the most cost effective solution. The cost of methanol is estimated at $6-7.5 million Btu's. However, coal derived liquids which have been severely hydrotreated to achieve nitrogen contents of less than 0.1 wt % are estimated to represent about the same cost as methanol.

TABLE VII

SRC-I Test Results (9)

Fuel	Fuel Analyses		Emissions			
			SO_2		NO_x	
	%S	%N	lb/10^6BTU	ppm	lb/10^6BTU	ppm
Coal	0.88	1.44	1.01	319	0.47	315
SRC-I	0.71	1.60	0.97	335	0.40	320

SRC-II Test Results (12)

Fuel	Fuel Analysis	NO_x Emissions	
	%N	Normal Boiler Setting	Low NO_x Boiler Setting
Petroleum Derived #6 Fuel Oil	0.23	155	100
Coal Derived SRC-II	1.00	270	175

Future Combustion Testing Programs

In 1979, because of a lack of large samples, boiler fuel test programs will be limited to small scale equipment. However, in 1980, large samples of liquids in the 5,000-10,000 barrel range should become availabel from the H-Coal pilot plant at Catlettsburg, Kentucky and the Exxon Donor Solvent process at Baytown, Texas. It would be preferable to run a number of tests utilizing different utility sites and types of electric generation equipment to allow several utilities to make a judgment as to what use these fuels may be to them and to establish acceptable safe handling procedures. Large scale utility test programs will require 10,000-40,000 barrels per day of fuel. Sustained test programs, that will last on the order of six months, must await successful operation of demonstration of pioneer commercial plants which are not scheduled to occur until after 1985.

The situation is somewhat different in terms of large scale combustion turbine test programs. Resumption of a methanol test burn is scheduled for early 1979. It was originally scheduled for a total of 500 hours of running time, averaging about four hours per day of actual operation. However a fire at the station, which was not related to the use of methanol, caused a six-month delay.

Although obviously not a coal liquefaction product, shale oil represents another synthetic fuel option. During the last quarter of 1979, the Department of Defense arranged with Standard Oil of Ohio through the Paraho Development Corporation to refine 100,000 barrels of raw shale oil. EPRI arranged for delivery of 4,500 barrels of the hydrotreated 700°F residue. This product will be used for a utility site combustion test during 1979.

Introduction of Coal Liquid to the Utility Market

It is not clear at this time how coal liquids will actually enter the utility market. One thing that is clear, however, is that products from the first demonstration or pioneer plants will not be competitive in cost with petroleum if these plants in fact are in production by 1985. This, taken with the utility industry's desire for extensive test programs prior to actual commercialization, makes this early "uneconomic" production of coal liquids a necessity if an orderly market is to develop. Therefore, some form of government action is required to provide a large supply of fuel for testing that will be required. We will leave the form of action to those more experienced in policy matters.

The larger question of what happens beyond the first few plants cannot be answered with any more certainty. Even the basic question of plant ownership offers a dilemma. Regulated utility financing would bring lower fuel costs to that utility. However, it means attempting to operate a complex facility without suitable corporate experience in refining and marketing. Energy company operation of a utility owned plant is another alternative. This offers a disadvantage to the energy company in that it must dedicate its people to such an endeavor for an uncertain market. Joint financing with energy company operations represents a possibility. The question of ownership is inevitably intertwined with that of plant product objective. If the plant produces a number of by-products, the owner must have the organization to market these by-products.

Another complication is that of product slate. An all distillate product would be compatible with petroleum liquids whereas a residual containing coal derived liquid would perhaps need to be segregated with dedicated storage and handling utility systems. As a result, these distillate products could be mixed with a non-dedicated product pool. Distillate products upgraded by hydrotreating would be even more acceptable products. A development strategy based on the marketing of high quality distillate products might be the easiest one to see through to successful commercialization.

Some consideration ought to be given to designing a first commercial or demonstration plant to maximize operability rather than profitability. This can perhaps be done by seeking out the areas of high process severity and backing off to milder operating conditions. For example, in each of the liquefaction processes that are considered to be relatively advanced, H-Coal, Exxon Donor Solvent, and SRC-II, reactors are run at high severities to maximize distillate yield. Then, in the case of the H-Coal and SRC-II processes all the vacuum tower residue is sent to a partial oxidation gasifier to produce hydrogen. The amount of residue is set by the amount of hydrogen to be generated. The Exxon Donor Solvent process differs in that all or part of the vacuum tower residue is processed in a Flexicoking unit to recover additional liquids and to produce low Btu fuel gas. Partial oxidation can be used to process the remainder of the bottom to produce hydrogen.

Plant configuration studies that maximize profitability seek to recover the maximum amount of distillate in the vacuum tower. This approach creates operability problems in both the hydrogenation reactor due to its high temperature and in the vacuum tower due to a solids loading of about fifty weight percent in the vacuum bottoms. It may be difficult to design a high reliability system to get this material out of the bottom of a vacuum tower because it has a high viscosity and high melting point.

The situation is further compounded when the gasifier or Flexicoker feed system is considered. Some surge capacity downstream of the vacuum tower is obviously required for good, steady plant operations. Unfortunately, vacuum tower bottoms are thermally unstable. Storage at high temperature causes its viscosity to increase. There are the obvious advantages to leaving the operability of the gasifier, Flexicoker and vacuum tower. The material that is sacrificed in a high boiling (800-1000°F) gas oil is solid at room temperature and contains more than 1.0% nitrogen.

A possible solution is to gasify the more dilute vacuum tower bottoms product in an oxygen blown gasifier and to convert the excess synthesis gas to methanol. In those cases where a Flexicoker is used the heavy scrubber liquids could be recycled to extinction. Therefore, the plant products are SNG, naphtha, 300-800°F distillate and methanol. All of these products are of high quality or can be hydrotreated to achieve high quality. As a result, they could be easily integrated into the utility fuel mix with a minimum amount of disruption or special product handling facilities.

This overall approach is a variation of the CDF process proposed originally by Lebowitz of EPRI. (15)

Summary

The production of clean solid and liquid fuels in the U.S. is on a path that leads to the production of significant quantities of synthetic fuels that are useful in power generation. Through the Electric Power Research Institute, the electricity industry has recognized its responsibility in providing support in the required research and development that is necessary. The Clean Liquid and Solid Fuels program area represents the largest annual expenditure of funds for a specific alternative technology. The program area has four basic elements that include:

- fundamental research
- support development of critical components
- process research in alternate routs to fuels, and
- definition of combustion practice in utilization of synthetic fuels.

This paper has primarily discussed the latter topic and other speakers at this conference have discussed a number of the other topics. It is likely that the large pilot plants that will begin operation in 1980 will establish engineering parameters and information that will bring the production of fuels from coal to technical readiness and provide a firm engineering and environmental data base to establish the foundation for a synthetic fuels industry in the U.S.

From an overall perspective, the operation of pilot plants in the 250-600 ton/day scale in the U.S. and in Germany will provide:

- engineering data and firmer product cost estimates
- environmental information useful for plant siting, and

o significant quantities of fuels for the electricity industry to test.

The next step of demonstration and pioneer plants from the standpoint of the utility industry is appropriate to provide 50-100,000 barrel quantities of these new fuels to complete the definition by the utility industry to transport, store, handle and utilize in electric generating equipment to generate power.

"Literature Cited"

1 "8th Annual Review of Overall Reliability and Adequacy of the North American Bulk Power Systems". National Electric Reliability Council, August 1978.

2 EPRI Technical Assessment Guide - May 1978.

3 Project Independence Report, Novemeber 1974, p. 122.

4 EPRI Report AF 741, "Process Engineering Evaluations of Alternate Coal Liquefaction Concepts". Final Report RP411-1, April 1978.

5 EPRI Report AF 710, "Economic Screening Evaluation of Upgrading Coal Liquids to Turbine Fuels". Final Report ITPS 76-666), March 1978.

6 EPRI Final Reports 1235-1, October 1975; 1235-3, August 1976; 1235-4, July 1976. "Investigating the Storage, Handling, and Combustion Characteristics of Solvent Refined Coal".

7 EPRI Final Report 1235-2a, June 1976. "Laboratory Analysis of Solvent Refined Coal-Technical Report #1". EPRI Final Report 1235-2b, June 1976. "Solvent Refined Coal Evaluation: Pulverization, Storage and Combustion-Technical Report #2".

8 EPRI Report FP628, January 1978. "Characteristics of Solvent Refined Coal: Dual Register Burner Tests". Final Report (RP1235-5).

9 "Solvent Refined Coal Burn Test" Southern Company Services, Inc., April 1978.

10 Muzio, L.J. and Arand, J.K. "Small Scale Evaluation of the Combustion and Emission Characteristics of SRC Oil" American Chemical Society Fuel Chemistry Symposium on Combustion of Coal and Synthetic Fuels, Anaheim, California, March 15, 1978.

11 EPRI Report FP1028, March 1979. "Characterization and Combustion of SRC-II Fuel Oil".

12 EPRI Report FP1029, March 1979. "Combustion Demonstration of SRC-II Fuel Oil in a Tangentially Fired Boiler".

13 Monthly Progress Reports - EPRI Research Project 989.

14 Pillsbury, P.W.; Cohn, A; Mulik, P.R.; Sihgh, P.P; Stein, T.R. "Fuel Effects in Recent Combustion Turbine Burner Tests of Six Coal Liquids". (Submitted for presentation to the ASME Gas Turbine Conference, March 11-15, 1979.)

15 EPRI Report EM622. "Clean Distillate Fuels Pilot Plant Study (Final Report RP916)".

CHAIRMAN SEGLIN: Thank you, Ron. We are one minute ahead of schedule. I will entertain one question.

SORAB R. VATCHA, Senior Research Engineer, Ashland Oil Co.: How can methanol at $6 or $7.50 per million Btu compete with intermediate Btu gas at about half the price?

R. WOLK: I think it's a question of how you deliver that intermediate Btu gas. We have a very small market in terms of Btu's for that service, and it has only been running maybe three to five-hundred hours a year at most. You can't afford to set up an intermediate gas plant for that kind of market.

ARTHUR L. CONN, President, Arthur L. Conn & Associates, Ltd.: You mentioned a great reduction in the use of gas, and I was wondering whether you have had a chance to react to this latest statement by the Department of Energy that there is more gas that can be used now and therefore, possibly there should be greater use of gas.

R. WOLK: I think I'll pass that question.

RECEIVED July 2, 1979.

2

Coal Dilemma II, "COGAS"

RALPH BLOOM, JR.

COGAS Development Company, P.O. Box 9, Princeton, NJ 08540

Based on the title of this symposium the objective of this paper is to discuss some dilemmas facing synthetic fuel process developers. The COGAS Process under development by the COGAS Development Company* is a combined liquefaction and gasification process. Development has been conducted since mid-1972 when the joint venture company was formed. We face two types of dilemmas.

* COGAS Development Company (CDC) is a partnership of:
- o Consolidated Gas Supply Corporation
- o FMC Corporation
- o Panhandle Eastern Pipe Line Company
- o Tennessee Gas Pipeline Company, a subsidiary of Tenneco, Inc.

Paraphrasing Shakespeare's Hamlet we could express the first dilemma as:

A synthetic fuels industry - to be or not to be

The second dilemma - competitive process economics are reported publicly on varying bases often with little detail.

Before discussing these two problems, the COGAS Process will be briefly described. If further detail is desired, CDC has available a number of papers.

The COGAS Process

The COGAS Process, Figure 1, features low-pressure conversion of coal to liquid products and high Btu substitute pipeline gas. The Process integrates multi-stage pyrolysis technology with steam gasification of char technology. Multi-stage pyrolysis was proven in a pilot plant of 36-tons-per-day of coal feed capacity which was operated successfully on a full range of coals from lignite through high-volatile A bituminous. Products of pyrolysis are oil, gas and low-volatile char.

0-8412-0516-7/79/47-110-023$05.00/0

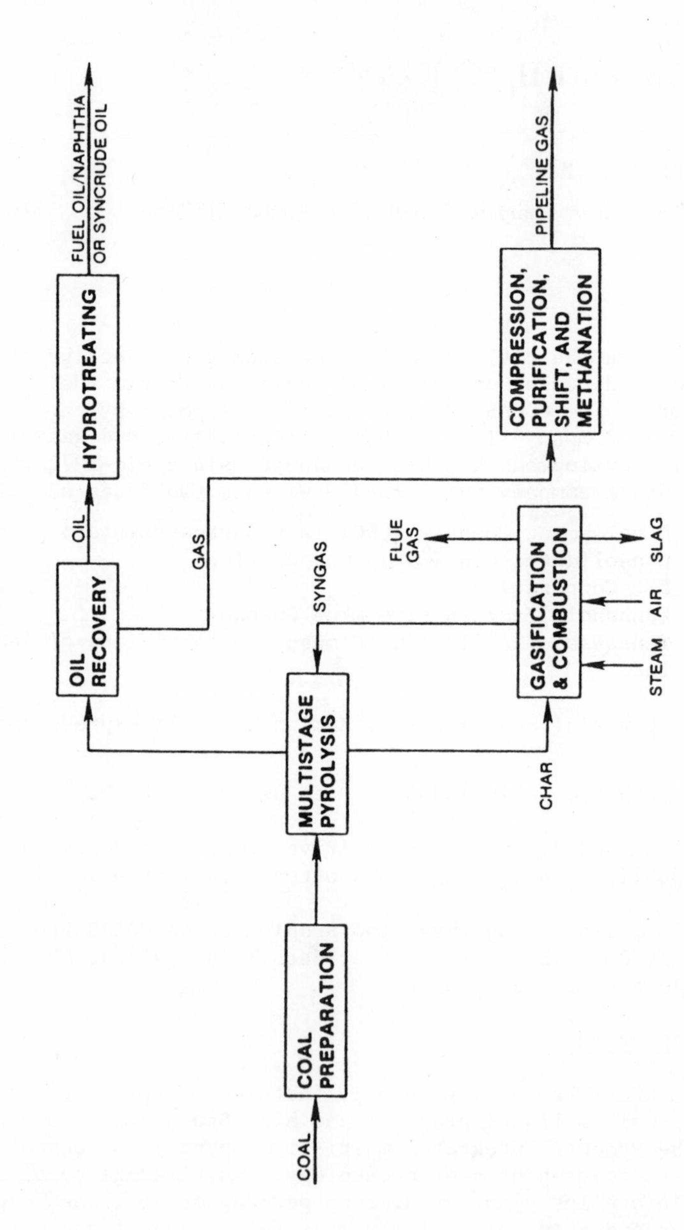

Figure 1. COGAS process

Promptly after formation of the COGAS Development Company, work started on the design and construction of a gasification pilot plant. Pilot-plant operation was initiated in March 1974. In addition, early in the program, process design engineering for commercial-scale plants was initiated. Cold models were also used effectively to develop the pilot-plant design and then to prove out elements of the commercial-scale design.

In the latter part of 1975, the development of the COGAS Process had proceeded to the point that it was considered ready for demonstration. On the basis of an extensive study and evaluation of second-generation coal gasification processes which were deemed to be ready or nearly ready for demonstration, the COGAS Process was selected by the Illinois Coal Gasification Group* (ICGG) for their proposal to the Energy Research and Development Administration (now Department of Energy, DOE) for the pipeline-gas-from-coal Demonstration Plant competition. This selection was based on the high thermal efficiency of the process for the production of synthetic pipeline gas and fuel oil and naphtha or synthetic crude oil. Also, the process had been piloted successfully on Illinois coal which was the primary coal for the ICGG Demonstration Plant.

*ICGG is a partnership of subsidiaries of five major Illinois gas utilities:

- o Northern Illinois Gas Company
- o The Peoples Gas Light and Coke Company
- o Central Illinois Public Service Company
- o Central Illinois Light Company
- o North Shore Gas Company

In June 1976, DOE selected the ICGG proposal as one of two proposals for contract. Work under DOE contract started in June 1977. The architect/engineer is the Dravo Corporation.

Continued development of the COGAS Process promises to help make our nation self-sufficient in meeting its needs for liquid and gaseous fuels. The process can handle all ranks of coals, ranging from lignite through high-volatile A bituminous coal. This versatility will be demonstrated further in the Demonstration Plant on three widely varying coal feeds.

The most recent conceptual commercial COGAS plant, Figure 2, produces 265 MM standard cubic feet of 950 Btu/scf pipeline gas per day, from bituminous coal plus 16,800 barrels per day of light (No. 4) fuel oil and 3800 barrels per day of gasoline reformer feedstock grade naphtha. Nitrogen content of this naphtha is less than 1 ppm. The combined gas and oil output from one such plant will permit a reduction of oil imports by as much as 22 MM bbl/yr. Coal feed rate is 26,000 tons per day or

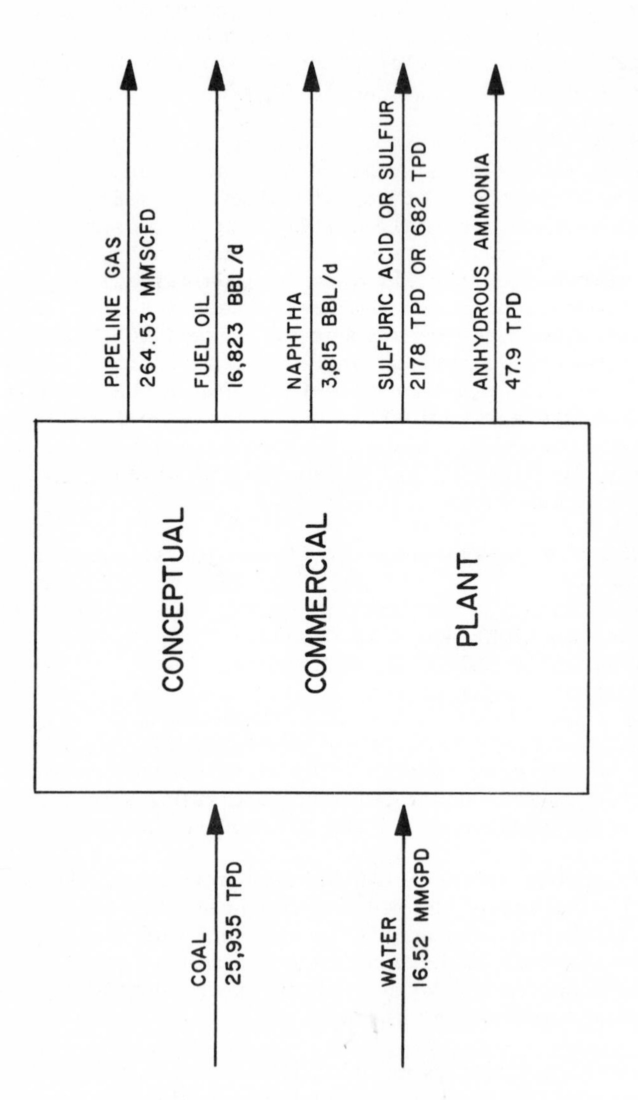

Figure 2. Conceptual commercial COGAS plant

8,600,000 tons per year based on 330 days per year on-stream time.

The COGAS Process promises to become an important means for this country to supplement its diminishing petroleum and natural gas supplies by the conversion of coal to clean-energy-fuels. Depending on continuing technical success, and a receptive economic climate, this promise should be achieved in the late 1980's.

Dilemma I

All the above sounds great, doesn't it? Our process development has proceeded successfully, initially with private financing by the CDC partners, more recently with Department of Energy financing. We are proceeding with the Demonstration Plant design program. Construction and operation is to follow, financed jointly by ICGG and the Government. But - will commercial plants ever be built using the COGAS Process or any other coal liquefaction or gasification process?

Much has been said in the past about the problems of generating a synthetic fuels industry - an industry which may require as many as 100 major plants in the 1990's (1) - not very long from now when you look at development, funding, siting, permitting and construction schedules. A very small sample of what has been said before includes Mr. A. C. Bellas' paper on Financing Coal Gasification Projects at the October 1975 Synthetic Pipeline Gas Symposium (2) and most of the papers and discussion at this Division's Excellent Symposium on Commercialization of Synthetic Fuels (3), three years ago. All the problems discussed in these two examples are still with us in 1979 and show no signs of going away. No projects have been started using existing, so-called "first generation" technology and the developing technology faces just as uncertain a commercialization future. The 1990's are steadily getting closer, - but the initiation of a synthetic fuels industry does not seem to be moving nearly as steadily.

I would like to cite a few specifics of the situation today. Using the COGAS Process as an example, the most recent estimate of the total plant investment cost of the commercial COGAS plant is $1.4 billion in mid-1978 dollars (4). In addition, there will be costs for land, adminstration during construction, start-up, working capital requirement to $1.5 billion exclusive of interest during construction before the plant produces at design capacity.

Continuing inflation will increase these costs further. For example, the design of a first COGAS commercial plant could be started in 1986 at the end of the second year of operation of the ICGG Demonstration Plant, assuming the program proceeds as scheduled with no further delays in decisions or financing. If capital costs escalate at 7% per year, the $1.4 billion plant investment estimate would increase to $2.4 billion in 1986 dollars. At this same average escalation rate this 1986 capital cost could increase by 50 percent over the design and construction period of about five years and the potential substantial additional time for obtaining authorizations and permits, fighting lawsuits, etc.

Certainly, there are not many corporations today that could afford - even if they had the assets - to put up their assets for such a plant. Financing would be a substantial problem because of the enormous investments, particularly for a process which has not previously been practiced commercially. Of course, we expect that operation of the COGAS Demonstration Plant will develop the confidence in the process that will be required for financing a commercial plant.

So, what's the answer - the U.S. Government? Maybe the balance-of-payments situation and its influence on inflation, plus the beginning of a worldwide oil shortage, will become serious enough to move the Congress and the Administration to take actions to make such investments possible. The forthcoming debate over the FY1980 budget may show the attitude of the U.S. toward preparing for such eventuality.

Look at the example of the Great Plains Coal Gasification* Phase I Project for producing 137.5 million standard cubic feet per day of synthetic pipeline quality gas from lignite via the Lurgi dry-bottom process, considered a commercially proven process because of its use in other countries since the late 1930's. To proceed with this project, approval was sought from the Federal Energy Regulatory Commission (F.E.R.C.) for surcharges and loan guarantees required to help finance the total of $904,488,000 in 1978 dollars estimated to be required for the project (5,6). The DOE was reported in June 1978 (7) to have advised the consortium that it would join in asking F.E.R.C. for orders providing:

1. Full recovery of debt capital plus interest if the project is abandoned. Advance approval of a tariff calling for system-wide rate payers to cover losses.
2. Initial assurance that 60% of equity would be recovered in the event of project non-completion and the right for investors to seek recovery of the remaining 40% in separate prochedings.
3. Current recovery of interest on debt during construction.
4. Rolled-in pricing for the coal gas at all levels and to all categories of customers.
5. Cost-of-service tariff for sale of pipeline quality gas by the partnership to pipeline members.

* Project Sponsors:
Great Plains Gasification Associates (American Natural Resources and Peoples Gas), Columbia Gas Transmission Co., Michigan Wisconsin PipeLine Co., Natural Gas Pipeline Co., of American, Tennessee Gas Pipeline Co., a division of Tenneco, Inc., Transcontinental Gas Pipeline Corp.

The DOE had announced that this was a synthetic fuel commercialization project it would strongly support. So what has happened? After public hearings, the F.E.R.C. staff filed a 24-page motion with the Administrative Law Judge to dismiss the case with prejudice. The principal problems in this case, are the high capital cost, and the high initial gas price and - as it will be in all synthetic gas cases - who will take the financial risk. And that case was only for production of 40 billion cubic feet of gas a year, 2/10 of one percent of the current U.S. consumption. (The U.S. consumption is about 20 trillion cubic feet a year).

The cost of synthetic fuels must be looked at in light of the years of production. If plant investments were made now the escalation effect over a 20-year production period would be reversed. For example, Great Plains showed, Figure 3, that with plant construction starting in 1978 the synthetic pipeline gas would initially cost substantially more than natural gas - but over a 20-year period it would be considerably less costly.

No corporation or consortium has yet sought to finance a commercial plant for producing liquids from coal; so we have no example to discuss, but we feel most of the same problems exist even though F.E.R.C. would not be involved.

Financing is probably the greatest constraint for the synthetic fuels industry, but there are others. Two examples are locating a site and obtaining the necessary permits and water supply. Recently it was reported (8) that 22 authorizations from 14 agencies are required for construction and operation of a synthetic pipeline gas plant.

Dilemma II

The second dilemma for a synthetic fuels process developer is related to "selling" the process. To be put to commercial use, the process under development must not only produce the products required, but must be shown to do so at costs that are competitive with other supplemental sources. The problem is to obtain economic analysis information on a consistent basis. A review of published economics indicates that it would probably be difficult to do this from papers presented at public meetings. Thus, for choosing a developing process to be used - or even to be supported

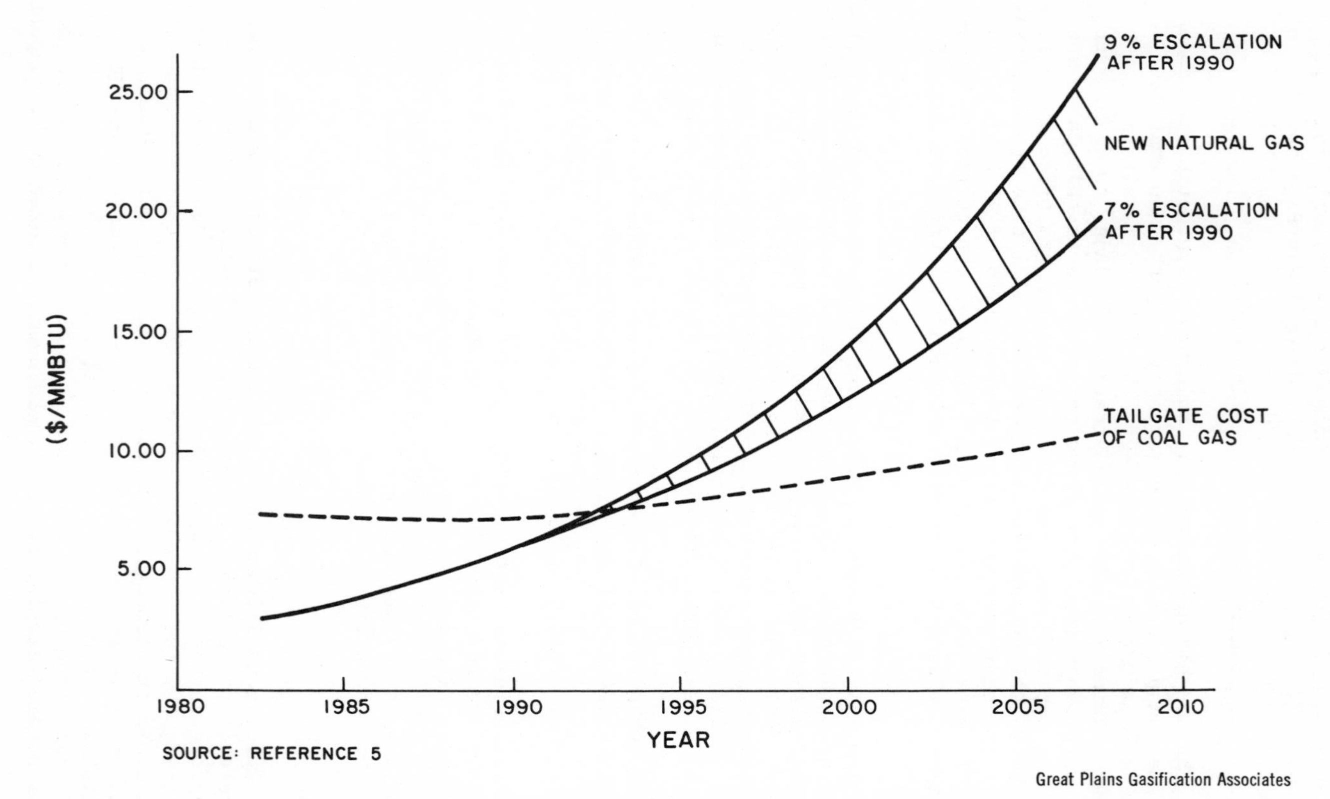

Figure 3. Great Plains Gasification Associates comparison of new natural gas wellhead prices to tailgate coal gas prices

it is necessary to have a study carried out which would put all processes on the same basis and provide an impartial analysis of how to apply the technology - an expensive study, if several processes are involved.

We cannot go into all the details here of the inputs into economic estimates, but the most important items which must be spelled out for meaningful interpretation are:

1. The degree of development of the design, the extent of vendor quotes and the contingency used in the capital estimates.
2. The cost data base used in the capital estimates; for example, cost estimaters of engineering firms which have build chemical process plants and refineries have available an extensive data bank from their experience.
3. The year in which the economics are based, including the escalation rates, if applicable.
4. The price of the coal delivered to the plant, and the basis for all of the cost elements of the operating cost estimate.
5. The way maintenance costs are estimated and the split of maintenance labor and materials.
6. The financial factors such as equity, debt, interest rates, depreciation, income tax rate, investment tax credit, entitlement, rate of return on equity and/or DCF rate.
7. The type of financing - utility or industrial - and, in the case of utility-type, whether the product price is first year or average over a specified period of years.
8. The quantities of products and by-products and the prices for the by-products.

Coal liquefaction analyses would be based on industrial financing, gasification projects for producing pipeline quality gas would be utility-type financing. In the case of a hybrid process such as COGAS which from a bituminous coal would produce about 65 percent gas and 35 percent liquids, on a Btu basis, we have generally used utility-type financing with the co-product liquids given by-product credit against production costs.

An example of the confusion that arises comes from the economic data presented at the Synthetic Pipeline Gas Symposium in October 1978.

C. F. Braun presented a fine reference paper (9) on gasification plant sizing using one process from their Western subbituminous coal study of processes as an example. A table was presented, dated Sept. 1978, which presented average 20-year gas costs in January 1976 dollars. C. F. Braun presented another paper (10) on their more recent Eastern bituminous coal study. This paper pointed out that certain changes in the procedure for computing operating costs were made which reduced the significance of a comparison of costs between the eastern and western coals. A number of excellent figures and tables of cost data were presented. Only one table, the detailed table of the capital cost estimate, noted in a footnote that the basis was also January 1976 while at the top of the table it was dated March 1978.

Capital cost estimates can, as noted above, be a problem. C. F. Braun stated that the data bases were such that design assumptions for the commercial concepts were not all confirmed and capital estimates might be optimistically low.

The C. F. Braun papers have been presented to summarize the results of studies which they have reported in detail in DOE reports. Their studies are the only ones that are available to the public which present economics for multiple high Btu gas processes on a consistent basis by one organization. But one must be careful in using the information in the papers to note the dates and the caveats.

Other papers also presented economic data, but not necessarily using the C. F. Braun economic guidelines. Three papers, (11,12,4), dealing with processes under consideration for demonstration plants sponsored by the DOE included economic information. In the paper by Procon on the HYGAS Process (11), all of the details were spelled out and gas costs were presented on four bases. Capital requirements are based on the conceptual commercial plant design and cost estimates done by Procon. The 20-year average gas price presented for bituminous coal by the utility financing method was \$3.78 with \$61.3MM by-product credit in 1978 dollars. For a similar plant, C. F. Braun figures were \$3.69/MMBtu with \$25MM by-product credits in 1976 dollars. Total plant investment capital costs were \$1,006,000,000 in 1978 dollars and \$930,000,000 in 1976 dollars respectively. The C. F. Braun plant was based on 250 billion Btu/day with no gas heating value specified while the Procon plant was based on producing 250MM scfd of 990 Btu/scf gas.

For the BGC/Lurgi Slagging Gasifier process (12) economic details for a conceptual commercial plant were not presented. The author stated that gas cost would be less than \$5/MMBtu on a utility-financing basis with 12% return on equity (13).

The COGAS Process (4) was presented by the senior author from the Illinois Coal Gasification Group, the prime contractor for the DOE demonstration plant program. Economics for the conceptual commercial plant were presented in mid-1978 dollars.

Plant investment was prepared by the Dravo Corp. Gas price was presented on the basis of "a typical utilities guidelines" which differed in many details from the utility financing method of C. F. Braun. In the case of COGAS, liquid product credit has a substantial effect on the gas price. In the paper this credit was at current market prices of $15.40/bbl for No. 4 fuel oil and $16.80 for naphtha. The resulting plant tailgate gas price on a 20-year operating time DCF basis was $5.08/MMBtu. However, if the liquids and gas are priced on an equivalent Btu basis, the fuel oil would be $25/bbl, the naphtha $27/bbl and the gas $4.10/MMBtu. These latter liquid prices are in the range of those estimated for liquids from coal by other processes.

For a so-called "advanced process" of flash hydropyrolysis, (14), a paper by Rockwell International and Cities Service Research and Development reported a 1977 minimum high Btu gas price of $2.36/MMBtu from western subbituminous coal using "AGA/ERDA cost guidelines" with utility financing under conditions yielding significant quantities of by-product BTX liquids. For details, reference was made to contractual reports.

When considering processes in early stages of development, such as the Rockwell process, one must consider the statement of Exxon in their paper on their catalytic coal gasification process: (15) "Exxon's experience in process development has shown that as a process moves through development the estimated cost invariably rises. To compensate for this historical trend we add contingencies to estimate the investment required for a first commercial plant". The amount of the contingency is a matter of judgement and will vary with the developer. CDC's experience is similar to that of Exxon. As detailed designs are developed, costs increase.

With varying economic information, such as discussed above, being presented at one meeting, it is no wonder that potential users of such processes might be confused as to which ones are the most attractive. However, the problem is not simple to resolve. Keeping conceptual commercial plant designs and economic analyses current with processes development is time-consuming and expensive. So when papers are presented, the authors have to use the data available. Thus, the process furthest along in development and with the latest economic analyses are liable to show the highest product cost.

The DOE attempt at standardized analyses as done by C. F. Braun is not the complete answer. Only the five processes in the DOE/AGA development program plus Lurgi dry-bottom were included and C. F. Braun's caveat on the capital cost estimates is significant since capital related costs are a substantial portion of the synthetic fuel product costs.

Conclusions

So synthetic fuel process developers have the two dilemmas discussed herein - when will there be a commercial synthetic fuesl industry and is the process under development going to be competitive. Hopefully, the Government will make the moves necessary to produce the investments in commercial-scale plants soon. COGAS Development Company feels it has the competitive process.

"Literature Cited"

1 McCormick, Wm. "Perspective on Synthetic Fuels", Symposium on Commercialization of Synthetic Fuels, Feb. 1976.

2 Bellas, A. C. "Financing Coal Gasification Projects", Seventh Synthetic Pipeline Gas Symposium, October, 1975.

3 Proceedings, Symposium on Commercialization of Synthetic Fuels, Division of Industrial & Engineering Chem., A.C.S., Feb., 1976

4 Eby, R. J., McClintock, N., Bloom, R. Jr., "The Illinois Coal Gasification Group Project - COGAS Process", 10th Synthetic Pipeline Gas Symposium, October, 1978.

5 Great Plains Gasification Associates, et al., "Additional Prepared Testimony of Eugene T. Zaborowski", July 7, 1978 and "Additional Prepared Testimony by Rodney E. Boulanger", July 14, 1978.

6 ANG Coal Gasification Co., Michigan Wisconsin Pipeline Co. "Additional Prepared Direct Testimony and Exhibits, Docket No. CP 75-278" Aug. 5, 1977.

7 Anon., "DOE Backs First Coal Gasification Plant", Oil and Gas Journal, June 12, 1978.

8 Dillon, R. E., and Newsom, H. R., "Commercialization of Coal Gasification". The National "Conference on the Impact of the NEA on Utilities and Industries Due to Conversion to Coal", December, 1978.

9 Maifield, D., Musgrove, R., "Considerations in Coal Gasification Plant Size", 10th Synthetic Pipeline Gas Symposium, Oct., 1978.

10 Detman, R., "Preliminary Estimates for Gasification of Eastern Coal", 10th Synthetic Pipeline Gas Symposium, Oct., 1978.

11 Vierk, H. S., "Conceptual Commercial HYGAS Plant Design", 10th Synthetic Pipeline Gas Symposium, Oct., 1978.

12 Verner, R. A. and Sudbury, J. D., "Slagging Coal Gasification in Industry and Government", 10th Synthetic Pipeline Gas Symposium, Oct., 1978.

13 Author's notes from 10th Synthetic Pipeline Gas Symposium, Oct., 1978.

14 Friedman, J., Combo, L. P., Silverman, Jr., Greene, M. I., "The Rockwell Advanced SNG Gasifier", 10th Synthetic Pipeline Gas Symposium, Oct., 1978.

15 Furlong, L. E. and Nahas, N. C., "Catalytic Coal Gasification Process Research and Development", 10th Synthetic Pipeline Gas Symposium, Oct., 1978.

RECEIVED July 23, 1979.

Hydrocarbonization

H. D. COCHRAN, JR.
Oak Ridge National Laboratory, TN 37830

Hydrocarbonization processes produce liquid, gaseous, and solid fuels from coal by low-temperature carbonization under hydrogen pressure. Hydrocarbonization is a relatively recent scion of the venerable class of low-temperature carbonization process, having been largely developed since World War II. This paper will review, generically, the effects of process variables on product yields, product quality, and hydrogen consumption. It will then present a brief historical overview of process development in the broad area of hydrocarbonization technology. This will lead to a general discussion of major process alternatives with reference to specific processes. Technological developments in problem areas for hydrocarbonization processes will then be described as background for an assessment of the present status and future prospects of this technology.

EFFECTS OF PROCESS VARIABLES

Hydrocarbonization processes are characterized by three primary independent variables - temperature, hydrogen pressure, and coal type - and five other, important independent variables - solid residence time, gas residence time, reactor configuration, coal pretreatment, and catalyst impregnation. Control of these variables permits control, over a wide range, of (1) the relative yields of liquid, gaseous, and solid products, (2) the quality of one or more of these products, (3) hydrogen consumption, and, ultimately (4) product cost.

0-8412-0516-7/79/47-110-037$05.00/0

Effects of Temperature, H_2 Pressure, and Coal Type on Yields

Among all independent variables, temperature has perhaps the most pronounced effect on yields from hydrocarbonization processes. Representative yields (1) from hydrocarbonization of Wyodak coal at a hydrogen pressure of 300 psi are shown in Figure 1. Typically, the yield of liquid products (oil and tars) shows a gentle maximum at a temperature about 1050°F. At higher temperatures, the maximum is reached when the liquid products are degraded to char and gas; thus, the temperature of maximum liquid yield may be shifted upward by reducing the time during which liquids are exposed to cracking conditions. Char yield decreases monotonically with increasing temperature as a result of increasing devolatilization and hydrogasification of the char. Gas yield increases monotonically with increasing temperatures, while water yield is relatively insensitive to temperature.

Figure 2 indicates the manner in which yields (2) from hydrocarbonization are influenced by hydrogen pressure. As expected, increased hydrogen pressure results in increased yields of liquid and gaseous products and, consequently, in decreased char yields. Generally, it is believed that hydrogen pressure increases liquid yields by stabilizing the radical fragments of initial pyrolytic decomposition in competition with parallel polymerization and cracking reactions which lead to loss of liquid products. Hydrogen pressure results in a small increase in water yields from ambient to moderate pressures (~300 psi), but the increase from moderate to high pressures (~1000 psi) is essentially negligible.

Little systematic, quantitative information is available concerning the effects of coal type on hydrocarbonization yields. In general, however, hydrocarbonization yields may be estimated by normalization of known results by the Fisher assay of the coal tested and thereby extended to other coals. The pronounced effects of coal type on operability and product quality are reviewd below.

Effects of Other Variables on Hydrocarbonization Yields

The primary devolatilization of coal is a very rapid, thermal process and therefore not strongly sensitive to solid residence

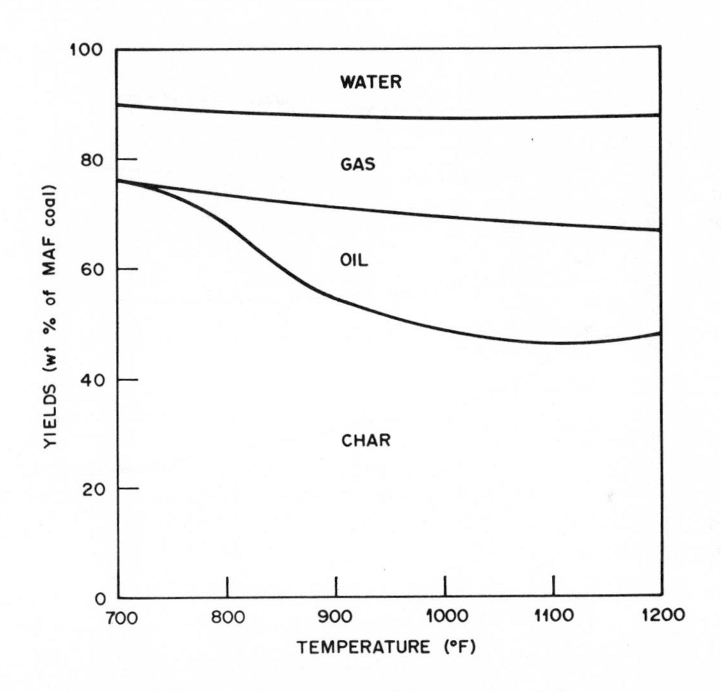

Figure 1. Hydrocarbonization yields for subbituminous coal at 300 psi H_2 pressure

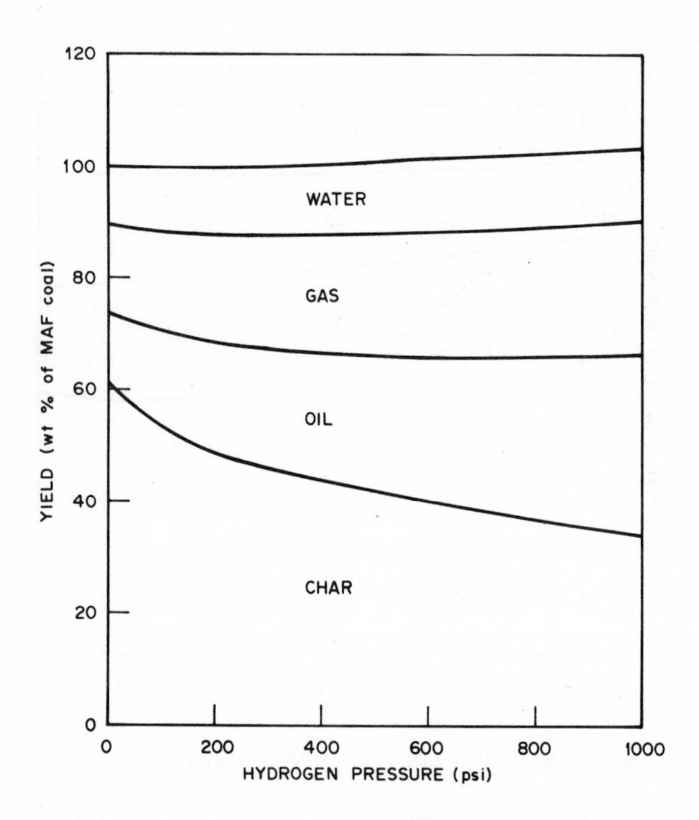

Figure 2. Hydrocarbonization yields for subbituminous coal at 1050°F

time. Secondary devolatilization and hydrogasification are slower processes, however, and result in increases of gas yield at the expense of char upon increased solid residence time. In contrast, the liquid products of hydrocarbonization are thermally unstable at processing conditions, so that increased liquid yields may be obtained with reduced char and gas yields by decreasing the residence time of vapors in the severe reactor environment. This effect (and apparently not rapid heatup) is the basis for the so-called flash pyrolysis and hydropyrolysis processes. Coal particle size contributes to this effect through hindrance of the escape of volatiles; therefore, reduced particle size also leads to increased recovery of liquids.

Hydrocarbonization yields may also be influenced by catalyst impregnation and coal pretreatment. For example, the yields of liquid and gasious products may be substantially increased by impregnation of the feed coal with $ZnCl_2$ or other catalysts. (3) In contrast, air exposure during coal preparation has a pronounced detrimental effect on liquid yields, as shown in Figure 3.

Reactor configuration may affect hydrocarbonization yields through its effect on residence time and, perhaps, on gas/solid mixing. Hydrocarbonization processes have been investigated in fixed-bed, stirred-bed, fluidized-bed, recirculating-bed, and entrained-bed reactors. The primary effect of reactor configuration is apparently the increase of liquid yields relative to gas and char yields as vapor residence time is reduced. However, recent results at ORNL, shown in Figure 4, indicate that, when the same feed coal, experimental system, and temperature/pressure conditions are used, only minor differences are observed in the fluidized-bed, recirculating-bed, and entrained-bed yields.

Effect of Process Variables on Product Quality

Quality of the liquid products is influenced by process variables. Generally, both the percentage of light oil (250 to 500°F) and the percentage of benzene, toluene, and xylenes in the total liquid product increase with increased hydrogen pressure and with increased reaction temperature, while the percentage of high-boiling asphaltenes and tars decreases. Similarly, increased temperature and pressure result in a beneficial increase in the hydrogen content of the liquids and a decrease in the heteroatom content of the liquids. These results are consistent with the increased hydrogen consumption at more severe conditions as discussed below.

In a similar way, the composition of product gas is influenced by process conditions. The percentage of carbon dioxide in the gas decreases with increasing temperature, hydrogen pressure, or solid residence time. The percentage of carbon

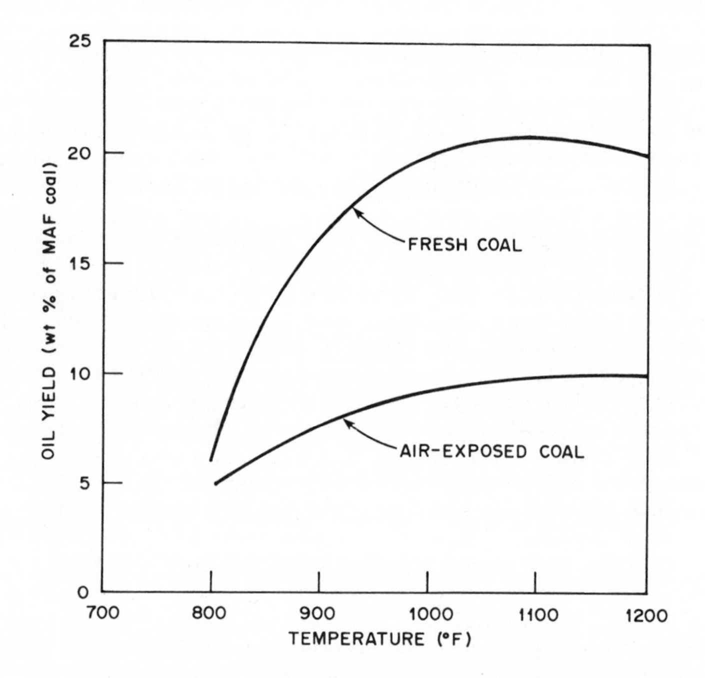

Figure 3. Effect of air exposure on oil yield for subbituminous coal

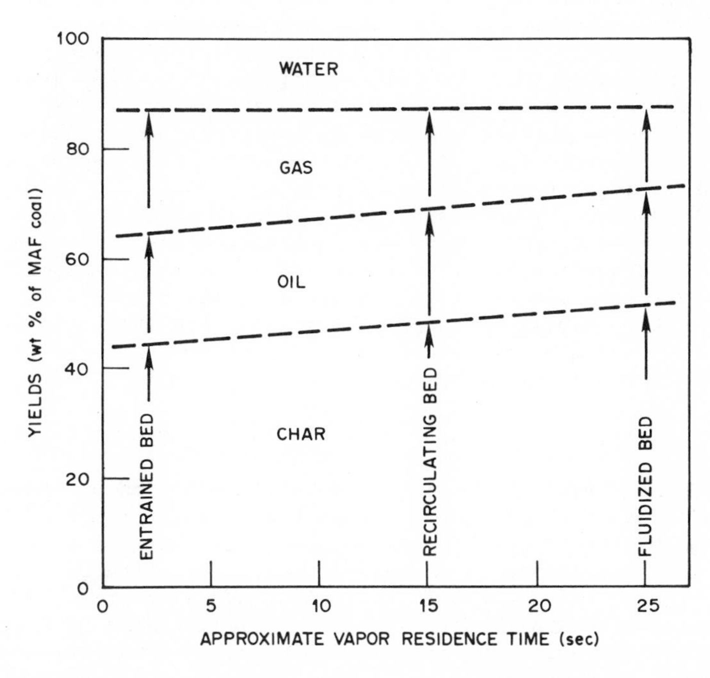

Figure 4. Effect of reactor configuration on yields (Wyodak coal; 300 psi; 1050°F)

monoxide and of methane in the gas increase with increasing temperature, hydrogen pressure, or solid residence time. The percentage of light hydrocarbon gases (C_2-C_4) also increases with severity of process conditions, but less markedly than that of methane (the relative increase varies roughly inversely with carbon number). These effects are typified by the results shown in Figure 5.

Char quality is best assessed by comparison with the coal from which it was produced. In this light, both its heating value and carbon content increase, while the volatile matter and sulfur content decrease with increasing severity of process conditions (increasing temperature, hydrogen pressure, or solid residence time). These trends are illustrated in Figure 6. It is of significance that the sulfur content (lb $SO_2/10^6$ Btu) can be substantially reduced by hydrocarbonization. Moreover, this reduction can be further enhanced by beneficiation of the coal prior to hydrocarbonization in order to produce low-sulfur char as a boiler fuel or metallurgical coke feedstock. In comparison to the feed coal, hydrocarbonization char is generally more reactive toward combustion or gasification because of its greater porosity and surface area. Further, it is of significance that chemical pretreatments, (1) which may be used to reduce agglomeration of caking coals, may employ alkaline salts which are retained in the char and are strong catalysts for steam gasification and methanation reactions. (4)

Effects of Process Variables on Hydrogen Consumption

Hydrogen consumption, in all coal liquefaction processes, is a variable of great practical importance because of the high cost of hydrogen generation. At ambient pressure there is a net generation of hydrogen from coal pyrolysis amounting to about 2 to 3 wt % of maf coal; with increasing hydrogen pressure, a net consumption of hydrogen occurs. This is illustrated in Figure 7, which shows the effect of reaction temperature as a parameter. It is a matter of practical significance that, at hydrogen pressures in the range 200 to 300 psi substantial quantities of coal are converted to liquid and gaseous products with no net consumption of hydrogen. Hydrogen consumption correlates directly with the degree of coal conversion and, therefore, with the reaction severity (temperature, hydrogen pressure, and solid residence time). Moreover, the exothermic heat of the hydrocarbonization process correlates well with hydrogen consumption; the heat of reaction per pound of hydrogen consumed decreases with increasing hydrogen consumption.

MAJOR PROCESS ALTERNATIVES

Figure 8 presents a brief historical overview of the develop-

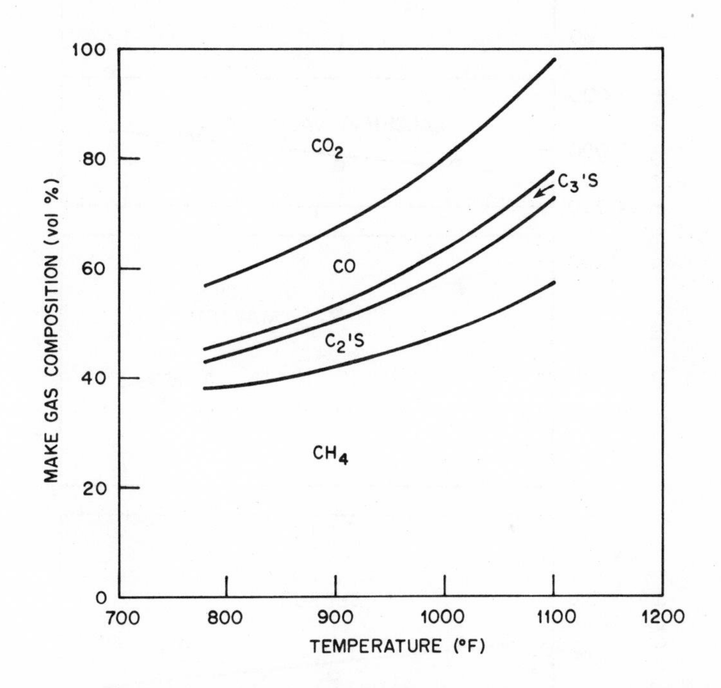

Figure 5. Effect of temperature on make gas composition for subbituminous coal at 300 psi

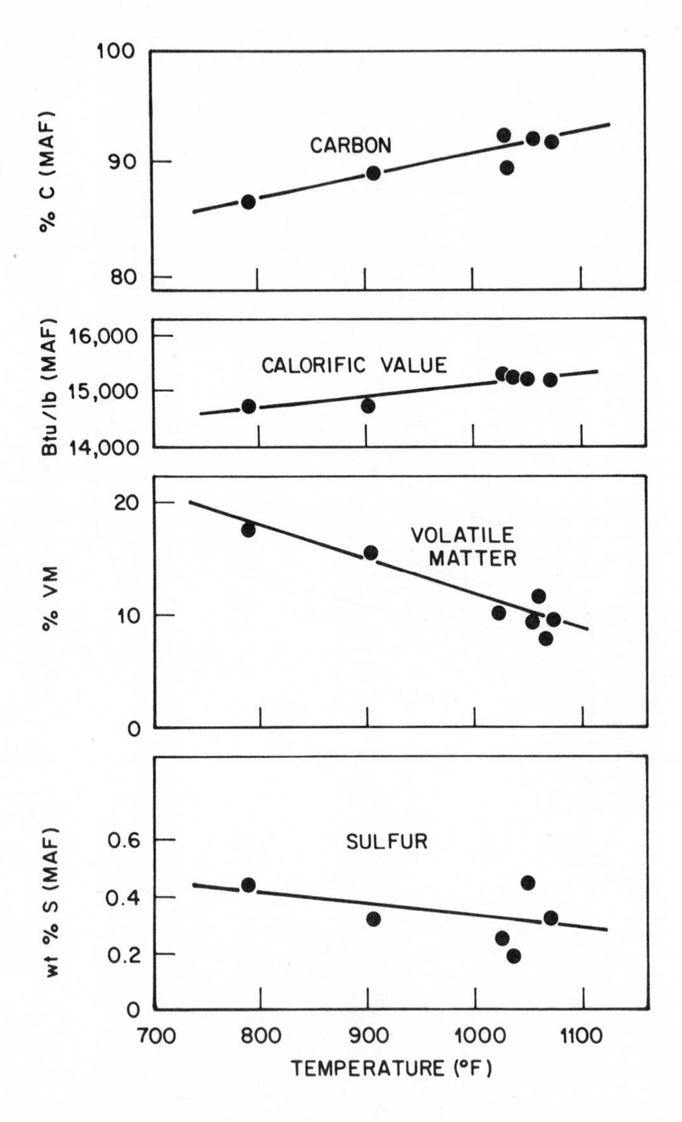

Figure 6. Properties of char produced from subbituminous coal at 300 psi

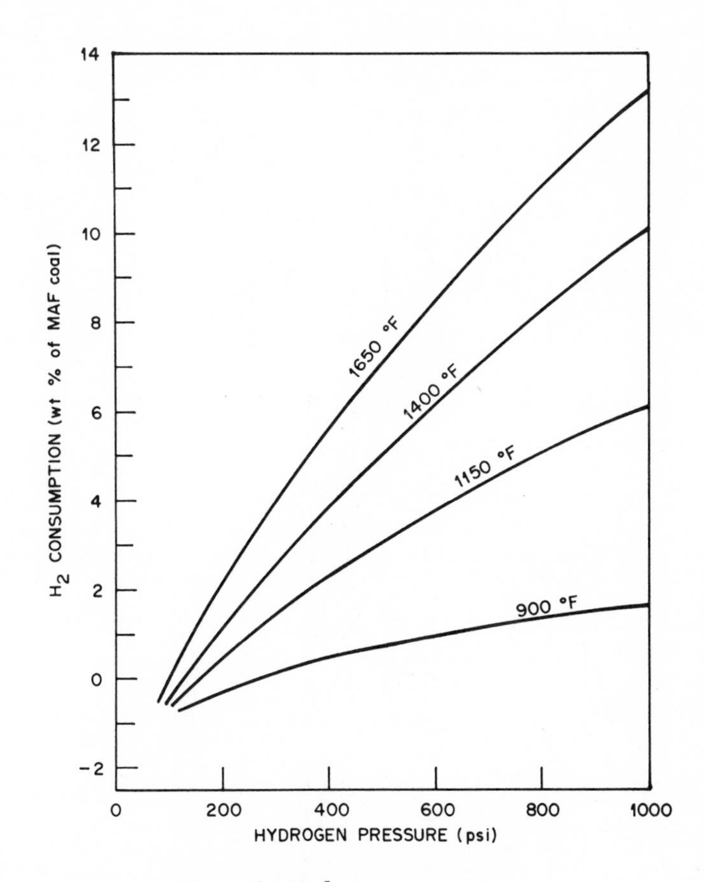

Figure 7. Hydrogen consumption

1943	Germany	--	Batch	Exploratory studies
1947-1954	Pittsburgh, PA	U. S. Bureau of Mines	Batch	Exploratory studies
1960-1965	Institute, WV	Union Carbide	Fluid-bed	18 TPD pilot plant
1960s	Library, PA	Consolidation Coal	Stirred-bed	1.5 TPD PDU
1969-1978	Monroeville, PA	U. S. Steel	Fluid-bed	0.5 TPD PDU
1975-1978	Oak Ridge, TN	Oak Ridge National Laboratory	Fluid-bed	Bench-scale
1975-present	Brookhaven, NY	Brookhaven National Laboratory	Entrained	Bench-scale
1976-present	Canoga Park, CA	Rocketdyne	Entrained	20 TPD PDU

Figure 8. Historical development of hydrocarbonization

ment of hydrocarbonization technology from exploratory studies in Germany during World War II, and further exploration by the U.S. Bureau of Mines during the 1940's and early 1950's, to the first substantial industrial developments through the mid-1960's. Interest in this technology was high in the early part of this decade but has lagged substantially since the failure of the Coalcon project, which was aimed at a large-sclae demonstration of hydrocarbonization technology. A few development activities continue in the United States, Great Britain, Australia, and perhaps elsewhere.

Because no single hydrocarbonization process is now the focus of attention, it is opportune to consider the major process options. Since hydrocarbonization yields three major products - liquids, gas, and char, there are at least three major process alternatives and several options of importance within each. These major process alternatives are listed in Figure 9, along with references to specific processes.

If the production of liquid products is to be optimized, two important alternatives must be considered: liquids production with no net char production and liquids production with no net hydrogen consumption. Both catalytic and non-catalytic examples of each are shown. In general, for the maximum production of liquids with no net char, it is necessary to operate at conditions of relatively high severity (e.g., $\sim 1050^{o}F$ and ≥ 550 psi). Therefore, the char product (≤ 37 wt %) is just sufficient to provide process needs for hydrogen (through gasification) and heat. This option was the basis for the Coalcon design. Alternatively, for maximum production of liquids with no net hydrogen consumption, milder conditions are appropriate (e.g., $\sim 1050^{o}F$ and 200 to 300 psi). Under such conditions, the char yield (~ 45 wt %) must be utilized either as a boiler fuel or as a gasifier feed stock. The ORNL mild hydrocarbonization process and the catalytic hydrocarbonization/gasification concept are examples of this alternative.

A combination of coal beneficiation and relatively high-temperature roasting of the char is required for production of low-sulfur char from high-sulfur coal. When an equilibrium recycle gas composition (at about 70 psi H_2) is used, char must be roasted at about $1400^{o}F$ for periods of about 1 hr, as in the U.S. Steel Clean Coke process. Alternatively, the use of low-sulfur coal permits production of low-sulfur char under a wider range of hydrocarbonization conditions so that higher liquid yields, for example, may be obtained.

Finally, a number of options exist for the production of high-Btu gas by hydrocarbonization and hydropyrolysis processes. In general, these processes involve operation at higher tempera-

I. PRODUCTION OF LIQUIDS
- A. WITH PRODUCTION OF NO NET CHAR
  - 1. NON-CATALYTIC
    - i. UNION CARBIDE/COALCON
    - ii. ROCKETDYNE
  - 2. CATALYTIC
    - i. SCHROEDER
    - ii. UNIVERSITY UTAH $ZnCl_2$
- B. WITH NO NET H_2 CONSUMPTION
  - 1. ORNL MILD HYDROCARBONIZATION
  - 2. (EXXON) CATALYTIC HYDROCARBONIZATION/GASIFICATION

II. PRODUCTION OF LOW SULFUR CHAR
- A. U. S. STEEL CLEAN COKE

III. PRODUCTION OF HIGH BTU GAS
- A. NON-CATALYTIC
  - 1. ROCKETDYNE
  - 2. HYDRANE
  - 3. COGAS
  - 4. HYGAS
- B. CATALYTIC
  - 1. EXXON CATALYTIC GASIFICATION
  - 2. SCHROEDER

Figure 9. Hydrocarbonization process alternatives, with examples

tures, 1500 to 1800°F, and may require higher pressures as in several of the rapid hydropyrolysis processes. The use of a catalyst permits high-Btu gas production at substantially milder conditions. For example, the methane net product of the Exxon catalytic gasification process is obtained at about 1300°F and 500 psi through recycle of hydrogen and carbon monoxide.

The available information leads one to believe that the maximum production of liquids with no net hydrogen consumption and the low-temperature catalytic hydrocarbonization/gasification are alternatives which appear to have great merit. The former of these, when applied to western coals, appears to be technically ready for commercial application and economically competitive with alternative coal liquefaction processes. Advantages of the flash hydropyrolysis processes over the Coalcon process are difficult to perceive.

PROBLEM AREAS IN HYDROCARBONIZATION

Hydrocarbonization processes suffer from problems that are uniquely associated with this technology as well as problems that are common to competing technologies. Paramount among those of a unique nature are the questions concerning char utilization and handling of caking coals. Problems common to hydrocarbonization and other coal conversion technologies include the feeding of dry solids to a pressurized system; the separation of gas, liquid, and solid products; the upgrading of products to marketable quality; and the optimal supply and utilization of process hydrogen and process heat.

Broadly, there are four acceptable approaches to utilization of hydrocarbonization char. If a low-sulfur char is produced, it may readily be used as a boiler fuel or as a feedstock for production of metallurgical coke. Alternatively, a high-sulfur char may be utilized as a boiler fuel either in a conventional furnace with flue gas desulfurization or in a fluidized-bed combustor. In general, utilization of high-sulfur char as a boiler fuel does not appear to be economically attractive. Char may be utilized as a gasifier feedstock; this possibility is particularly attractive when the char contains gasification catalyst used as a coal pretreatment prior to hydrocarbonization. Finally, as noted above, it is possible to optimize hydrocarbonization processes for the production of no net char.

Handling of caking coals has proved to be a serious obstacle to the development of hydrocarbonization processes and was, in fact, one of the principal factors contributing to the failure of the Coalcon project. However, a number of technologically successful approaches to handling of caking coals have now been demonstrated. The most common approach is through special reactor configurations. Typical examples of this approach include the COED multistage pyrolysis system, the Westinghouse recirculating bed, entrained flow reactors of the Rocketdyne type, and a proprietary reactor design demonstrated by Union Carbide after termination of the Coalcon project. An alternative approach involves chemical pretreatment of the coal. Preoxidation of the coal is technically feasible, but this pretreatment seriously reduces liquid production. Other approaches include the Battelle CaO-NaOH pretreatment, the Exxon KCO_3 or KOH pretreatment, and several other chemical pretreatments tested by ORNL. (1) Of these, at least the alkali salt pretreatments show positive advantages in other aspects of the process. Finally, one should keep in mind that substantial reserves of noncaking coals exist in the northern great plains and mountan provinces.

Solutions available to the problem of feeding of dry solids to pressurized systems include conventional lock hoppers, feeding of the coal as a slurry in light oil or BTX, and several advanced feeder concepts currently under development. Technologies for separation of hydrocarbonization product phases are similar to those employed in other liquefaction processes, with hydrocarbonization having the advantage of far lower solids content in the product liquids when high efficiency cyclones are used for char/vapor disengagement in the reactor. Procedures for upgrading the quality of hydrocarbonization products are also similar to the ones used in other liquefaction processes such as hydrotreating liquid products. If desired, the heavier fractions of the hydrocarbonization product liquid may be recycled to estinction in the hydrocarbonization reactor. Optimization of the generation and utilization of process hydrogen and process heat is a design exercise common to all liquefaction processes.

PRESENT STATUS AND FUTURE PROSPECTS OF HYDROCARBONIZATION

Presently, interest in hydrocarbonization technology appears to be at a low ebb, particularly in comparison with the high level of activity in the area of slurry hydroliquefaction technology. The failure of the Coalcon project has seemingly cast a pall over all hydrocarbonization development activities. The U.S. Department of Energy (DOE) is continuing to fund a small research project at Brookhaven National Laboratory and a larger development project on flash hydropyrolysis under Rocketdyne's leadership. Finally, the COGAS project, which is more correctly characterized as a pyrolysis/gasification project, is still a contender (with the slagging Lurgi gasifier) for this nation's first large demonstration of high-Btu gasification. The DOE's current lack of interest in hydrocarbonization technology seems to reflect a lack of confidence in its potential by private industry. Whether this is a correct appraisal of the situation remains to be seen.

The future prospects for hydrocarbonization technology are difficult to project, for, without support from the federal government, none of the technologies for producing liquid fuels from coal can compete with the current world price of petroleum. It should be kept in mind, however, that, at least for application to western coal, hydrocarbonization is a technically viable process which co-ld be commercialized with minimal technical risk. Moreover, it appears that hydrocarbonization processes are economically competitive with other coal liquefaction processes, at least within the range of uncertainty of available cost projections. Finally, it appears that current technological developments have successfully improved methods for addressing the problem associated with hydrocarbonization in a fashion that would appear to be to the competitive advantage of this liquefaction technology.

What, then, does the future hold? This author believes that the catalytic hydrocarbonization/gasification concept will ultimately achieve commercial success for the production of liquid and gaseous fuels from coal. In selected applications, the mild hydrocarbonization of western coal to produce liquid and gaseous fuels with power generation from the low-sulfur char may also be commercially attractive. Finally, further development of the flash hydropyrolysis technology, as exemplified by the Rocketdyne project, may eventually lead to a technically and economically attractive liquefaction process. But the most important questions still remain unanswered. Does private industry have sufficient interest to pursue the possibilities? Where is the interest focused? Will a private consortium build a hydrocarbonization/cogeneration complex using western coal? Will the phoenix arise from the ashes?

ABSTRACT

Hydrocarbonization, or low-temperature carbonization under hydrogen pressure, is representative of a class of coal conversion processes distinctly different from the slurry hydroliquefaction processes and processes which synthesize liquid fuels from co l-derived synthesis gas. Hydrocarbonization technology is reviewed, and major process alternatives and problem areas are discussed. The present status and future prospects for hydrocarbonization are assessed.

"Literature Cited"

1 H. D. Cochran and E. L. Youngblood, Hydrocarbonization Research: Completion Report, ORNL/TM-6693, in preparation.

2 J. M. Holmes et al., "Evaluation of Coal Carbonization Processes," in Coal Processing Technology, 3, AIChE, New York, 1977.

3 R. E. Wood and W. H. Wiser, "Coal Liquefaction in Coiled Tube Reactors, " Ind. Eng. Chem. Process Des. Dev. 15(1), 144 (1975).

4 W. R. Epperly and H. M. Siegel, "Catalytic Coal Gasification for SNG Production," Proc. 11th IECE Conf., Stateline, Nevada, 1976.

Research sponsored by the Fossil Energy Office, U.S. Department of Energy under contract W-7405-eng-26 with the Union Carbide Corporation.

RECEIVED July 2, 1979.

4

Production of Distillate Fuels by SRC-II

D. M. JACKSON and B. K. SCHMID

Gulf Mineral Resources Company, 1720 So. Belaire, Denver, CO 80222

The SRC-II process technology for the production of low-sulfur distillates and light hydrocarbons from coal has been tested and evaluated in laboratory and pilot plant experiments on a variety of high-sulfur coals. Its development has successfully evolved to the point where large scale demonstration of the process and required equipment can be considered. Gulf, through its Pittsburg & Midway Coal Mining Co. Subsidiary, is completing, under contract to the Department of Energy, a preliminary evaluation of engineering design, site, and market and economic assessment of an SRC-II demonstration plant. The facility will be located on a site suitable for a subsequent commercial facility near Morgantown, West Virginia. The feed coal for the demonstration plant will be a typical high-sulfur Pittsburgh seam coal from West Virginia.

The plant will yield significant quantities of coal liquids, gas and other products for extensive longer term testing in boilers, turbines and other applications.

The objectives of the demonstration program are:

1. To verify the technical feasibility of the SRC-II process in full-size equipment and establish a design basis for future plants.

2. To integrate various supporting processes such as high-pressure gasification into an overall coal liquefaction process.

3. To make production quantities of low-sulfur fuel oil, gaseous hydrocarbons and chemical by-products for longer term testing.

0-8412-0516-7/79/47-110-055$05.00/0

4. To develop appropriate systems and equipment for controlling any environmental, health, and safety factors that may be unique to large scale coal liquefaction plants and their products.

5. To provide a firm basis for estimating capital and operating costs required for a commercial coal refinery utilizing the SRC-II process.

PROCESS DESCRIPTION

Flow Scheme

Figure 1 presents a schematic flow diagram of the process in a full-scale plant as has been generally described in earlier publications (1,2,3). The feed coal is initially dried to about 5 percent moisture and pulverized, then mixed with recycle slurry from the process. The resulting coal-slurry mixture is pumped, together with hydrogen, through a fired preheater to a reactor at elevated temperature and pressure. In the reaction system the coal is not only dissolved, but is also largely hydrocracked to distillate fuel oil, naphtha and light hydrocarbons.

The reactor effluent then flows through a series of vapor-liquid separators, where it is separated into process gas, light hydrocarbon liquid and product slurry. The gas, consisting primarily of hydrogen and gaseous hydrocarbons, together with minor amounts of H_2S and CO_2, first goes through an acid gas removal step for removal of the H_2S and CO_2. The treated gas then goes to a cryogenic separation step for removal of the hydrocarbons. The purified hydrogen is recycled to the process, while the recovered hydrocarbons become by-products of the process. The C_1 fraction is sent to a methanator to convert the remaining CO to methane. The other light hydrocarbons are fractionated to produce ethane, propane and a mixed butane stream. The light hydrocarbon liquid goes to a fractionator where it is separated into naphtha (C_5-350°F nominal boiling range) and a middle distillate (350° - 600°F boiling range).

The product slurry is split, with one portion being recycled to the process for slurrying with the feed coal. The other portion of the product slurry goes to a vacuum tower where a heavy distillate is removed overhead. The heavy distillate, together with middle distillate from the fractionation step, makes up the total fuel oil product of the process.

The residue from the vacuum tower is sent to a high pressure slagging gasifier for production of synthesis gas. A portion of the synthesis gas goes through shift conversion and acid gas removal steps to produce pure hydrogen for the process. The

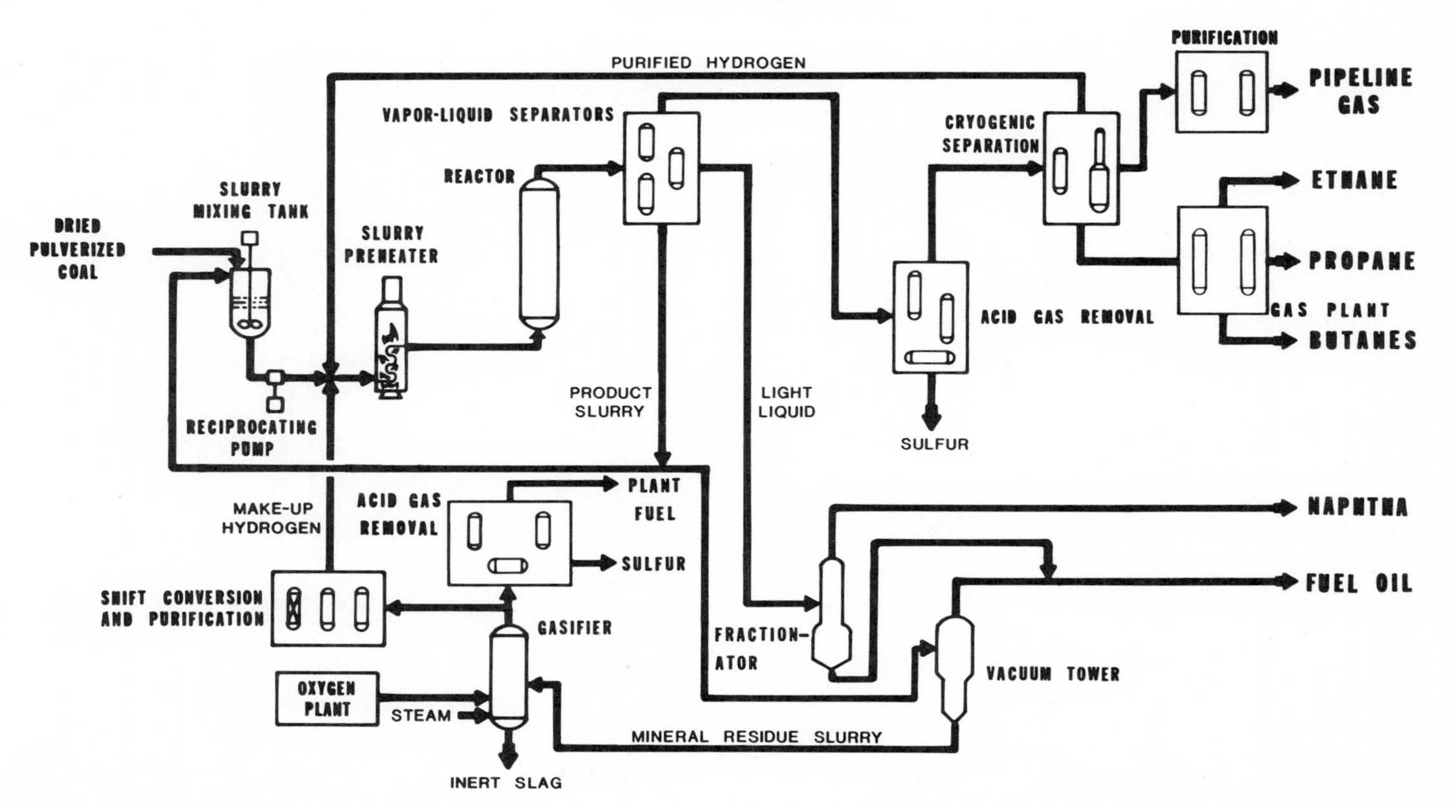

Figure 1. SRC-II process

synthesis gas in excess of that required for hydrogen production is passed through a separate acid gas removal step for removal of CO_2 and H_2S, then through a power recovery turbine, and is finally burned as plant fuel.

Major Process Steps and Related Engineering Development

The demonstration plant is expected to confirm the operability and reliability of those process steps and certain process equipment which have not yet been proven in commercial scale equipment in the operating environment of coal refining. Certain aspects of the engineering development of these areas are discussed, as shown in Table 1.

TABLE I

MAJOR SYSTEMS TO BE DEMONSTRATED

Slurry Mixing and Pumping

Slurry Preheater

Dissolver

Fractionation

Heat Exchange

Pressure Letdown

Gasification

Oxygen Compression

Slurry Mixing and Pumping

The demonstration plant will utilize a slurry mixing and pumping system which has appeared very promising in tests at the 50-ton per day pilot plant at Ft. Lewis, Washington. Coal is initially contacted with the slurry in a small mixing vessel to accomplish the initial mixing required for completely wetting the coal particles. Most of the 5% moisture remaining in the feed coal is vaporized in the mixing tank. The resulting thick slurry is then pumped to the main slurry mixing vessel where mixing is completed. The mixing step is complicated by the fact that the coal-slurry mixture forms a gel, and the rate of formation of the gel is strongly independent upon temperature. The formation of the gel greatly increases the viscosity of the mixture and makes mixing and pumping more difficult. Although the effect of the higher viscosity can be at least partially overcome by strong shear forces generated by appropriate mixers and pumps, these effects must be demonstrated in larger equipment.

Slurry Preheater

Similarly, the formation of the gel and its complicating effect upon the viscosity of the three-phase slurry mixture must be carefully managed in the slurry preheater. Measurements of pressure drop and heat transfer in the Ft. Lewis pilot plant have provided much valuable information concerning the effect of viscosity of the mixtures. For example, the observed pressure drop is significantly lower than would be calculated based on the viscosity estimated from laboratory test studies. This appears to result from the non-uniform temperature-viscosity gradient over the cross-section of the heater tube in the region where the gel is a significant factor in the viscosity. After the gel reaches its peak viscosity, the viscosity decreases rapidly as solvation proceeds. Thus, the gel nearest the hot wall is probably in a more advanced state of depolymerization and the viscosity of the fluid near the wall is significantly lower for much of the length of the preheater coil than the bulk fluid viscosity at the same cross-section. Even with the reduced pressure drop, however, the maximum practical tube diameter is limited by heat transfer, and this requires that multiple tube passes be used and proven in the demonstration plant.

Dissolvers

The basic design for the dissolver is a vertical pressure vessel with no internals. Continuing studies confirm that the reactor is well backmixed and that temperature should be reasonably uniform throughout the vessel, even in larger scale equipment. The highly exothermic hydrocracking reactions occurring in the dissolver make it feasible to feed the reactants at a temperature well below that prevailing in the dissolver. The effectiveness of hydrogen quench in controlling the reaction temperature has been confirmed in pilot plant tests and this technique will be employed in the larger demonstration plant vessel. The hydrogen quench is added at various points in the reactor and assists in maintaining the backmixing as well as serving as a fine temperature control.

Fractionation

Continuing study of the fractionation system for the SRC-II process, both in pilot plant and engineering work, has indicated that some modification to the original fractionation system design is desirable. In the original design the slurry was passed through the fractionator, then to a vacuum tower. In the revised design, however, the slurry bypasses the fractionator. Bypassing the fractionator has been made possible by more extensive flashing of lighter liquid from the slurry, thereby eliminating a difficult solids-handling problem in the fractionation step. The fractionator handles essentially all of the distillate liquids flashed during pressure letdown of the slurry, and separates the combined liquid into naphtha and middle distillate.

Heat Exchange

The first vapor-liquid separator following the dissolver separates excess hydrogen and uncondensed hydrocarbons from the product slurry. The vapor stream must then be cooled to condense normally liquid hydrocarbons. This cooling is carried out in a series of cooling and vapor-liquid separation steps, the first of which is a hot high-pressure heat exchanger. This exchanger requires careful design because of the probability that some solids carry-over may occur in the first separator, leading to the presence of solids in the exchanger. A major engineering effort was made to accomplish a design which should satisfactorily handle concurrently the problems of high temperature, high pressure, the presence of hydrogen and the presence of solids.

Pressure Letdown

The letdown of the hot slurry to lower pressures is also of concern because of potential erosion of letdown valves. The high velocity created by flashing vapors, combined with the presence of erosive solids, make this an important consideration in the mechanical design of the demonstration plant. Extensive studies have been carried out in the 50 ton per day pilot plant at Fort Lewis, and several arrangements and type of valves have been tested. This experience has led to design of a three-stage letdown system for the slurry in the demonstration plant. Testing of promising valve systems is continuing in the pilot plant.

Oxygen Compression

The design for the oxygen plant includes large centrifugal compressors for raising the oxygen pressure to the level required for the gasification step. Centrifugal compressors have been successfully operated in commercial installations at high pressure but now quite as high as the design pressure. A major engineering study, undertaken in consultation with oxygen compressor manufacturers, concludes that operation at the higher pressure appears feasible by the use of three casings of several stages each.

High-Pressure Gasification

High-pressure gasification of the vacuum bottoms permits thermally-efficient production of hydrogen from gasifying the carbonaceous matter in the mineral residue, as well as recovery of the inorganic matter as a relatively clean inert slag. High pressure operation of the slagging gasifier with the high solids content feed is an important element in the demonstration program.

PRODUCTS

Yields and Applications

Although the SRC-II process has been developed primarily for conversion of coal into distillate fuel oils, a number of other lighter hydrocarbon products are also obtained. The demonstration plant would be designed to produce primarily utility fuels for direct use without further refining and to permit product purchase support of the project by the utility industry. A subsequent commercial facility, while still producing significant quantities of fuels for boilers and turbines, offers the economies of scale for recovery and upgrading (as appropriate) of lighter hydrocarbons, as well as more selective product applications based on distillate product characteristics and end-use requirements.

A brief outline of the products expected in a demonstration plant and in future commercial plants is shown in Figure 2. In future commercial plants, for example, ethane and propane could be utilized as chemical intermediates and naphtha as a source of chemicals or for production of high-octane unleaded gasoline. Synthesis gas produced in excess of the requirements for hydrogen could be utilized as a source of chemicals as well as a fuel. The fuel oil could be selectively fractionated to produce a middle distillate for use as turbine fuel, light industrial boiler fuel or refinery feedstocks, while the heavy distillate could serve as a fuel oil for large utility boilers.

The anticipated product slate from a typical commercial plant feeding 33,500 tons per stream day of dry coal is given in Table II. This product slate is based on conversion of a typical Pittsburgh seam coal from West Virginia. The ultimate analysis of the coal used as a design basis is given in Table III.

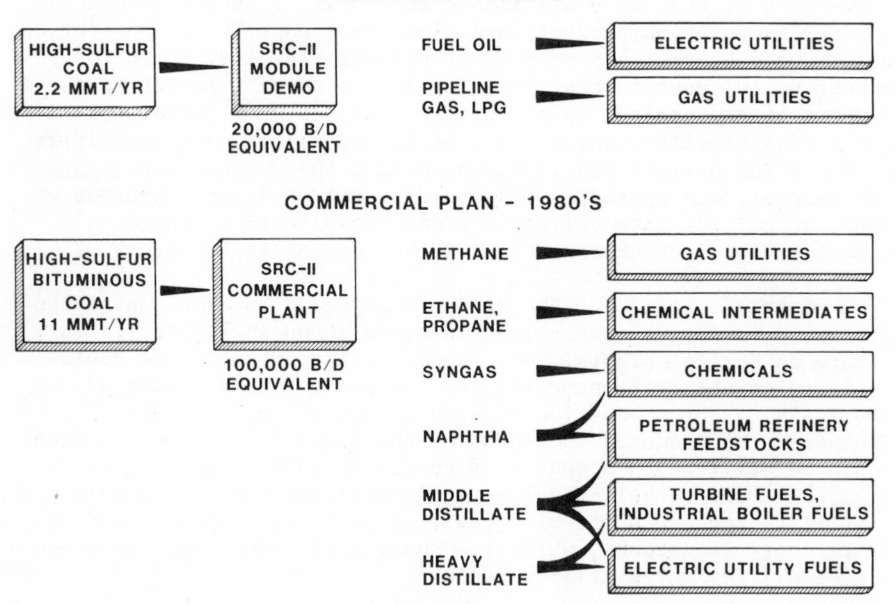

Figure 2. SRC-II product development

TABLE II

PRODUCTS FROM TYPICAL COMMERCIAL PLANT
33,500 T/SD-HIGH SULFUR BITUMINOUS COAL

METHANE	120 MM SCF/D
ETHANE	1,100 T/D
PROPANE	12,000 B/D
BUTANES	8,000 B/D
NAPHTHA (C_5-350^oF)	13,200 B/D
FUEL OIL (350-900^oF)	57,500 B/D
SULFUR	800 T/D
AMMONIA	150 T/D
PHENOLS	35 T/D

TABLE III

ANALYSIS OF FEED COAL

HIGH SULFUR BITUMINOUS COAL - PITTSBURGH SEAM

	% BY WT.
CARBON	71.0
HYDROGEN	5.0
NITROGEN	1.4
SULFUR, PYRITIC	1.6
SULFUR, ORGANIC	1.0
OXYGEN	7.0
ASH	12.0
MOISTURE	1.0
	100.0

The major market for the product fuel oil for the demonstration plant and near-term future commercial plants is expected to be existing power plants in the coastal metropolitan areas, where the physical and environmental costs of conversion to coal make such a conversion impractical. A significant characteristic of the SRC-II fuel oil for this application is its low sulfur content and thus the capability to meet stringent emission limits in urban areas. Coal-derived residual fuels will, in general, not meet these requirements without stack gas cleanup.

TABLE IV

PROPERTIES OF TEST FUELS

(Based on average analysis of samples taken during test program)

	No. 6 Fuel Oil	SRC-II Fuel Oil
Gravity: ^{o}API	25.0	11.0
Viscosity:		
SUS at 100°F	-	40
SUS at 122°F	300-700	-
Ultimate Analysis (Dry): % By Wt.		
Carbon	87.02	85.50
Hydrogen	12.49	8.86
Nitrogen	0.23	1.02
Sulfur	0.24	0.22
Oxygen	-	4.38
Ash	0.02	0.02
Heating Value: BTU/LB.	19.200	17.081

Table IV gives the properties of the SRC-II fuel oil compared to a low-sulfur residual oil utilized in a recent combustion test. The SRC-II fuel oil is a distillate product with a nominal boiling range of 350-900°F, a viscosity of 40 Saybolt seconds at 100°F and a pour point below -20°F. Thus, it is readily pumpable at all temperatures normally encountered in transportation of the fuel oil. The fuel oil has a very low content of ash and sediment as well as a low Conradson carbon residue. These characteristics are favorable from the standpoint of particulate emissions during combustion. Tests of compatibility with typical petroleum fuel oils and on stability of the coal distillates over time have not revealed any unusual characteristics that would preclude utilization of these coal-derived fuels in conventional boiler applications.

Combustion Characteristics

The major question involving burning characteristics of coal liquids relates to the higher nitrogen content compared to petroleum fuel oils and the potential effect on NO_x emissions. Since NO_x emissions are sensitive to burning conditions, however, actual burning tests are required under various conditions to assess the effects.

Several burning test programs have been carried out to confirm that the SRC-II fuel oil could be successfully used in conventional power plants and that emission levels of potential atmospheric contaminants could be controlled.

The first burning test for the liquid fuel oil was conducted in a 3 MM Btu per hour test boiler. The fuel handling characteristics of the oil were attractive. Viscosity was comparable to No. 2 fuel oil, thus no preheating was required. The SRC fuel oil was used interchangeably with No. 2 fuel oil without forming sediments. Cold boiler light-offs were made without incident. Although the fuel oil has the relatively high organic nitrogen content characteristic of coal-derived liquids, each of several combustion control technologies were effective in decreasing NO_x formation and smoke to environmentally acceptable levels. These combustion control methods include staged combustion, steam atomization, low-NO_x burner design, and smoke inhibiting additives.

In the fall of 1978 a full-scale test program was pursued in a commercial power plant of the Consolidated Edison Company in New York City (4). The test was conducted in three phases in Con Edison's 74th street station utilizing a 450,000 lb/hr steam electric Combustion Engineering tangentially-fired boiler, as shown in Table V.

PHASE I - Initial Baseline Testing

Work in the first phase involved preliminary checking of equipment and instruments for measuring emissions, as well as establishment of NO_x reduction trends using staged combustion techniques, while burning the current power plant fuel, a low-sulfur No. 6 fuel oil. The purpose of this phase was to reduce the time necessary to carry out the subsequent SRC-II tests and to achieve minimum NO_x levels with the limited supply (4,500 bbls) of SRC-II fuel oil.

PHASE II - SRC-II Fuel Oil Testing

The second phase involved a 6-day test of the SRC-II fuel oil to determine its combustion performance and emission levels under various operating conditions. Tests were made at full load, three-quarter load and one-half load while using normal combustion (baseline) and staged combustion techniques. The staged combustion tests were made to evaluate the possibility for substantially decreasing NO_x emission levels.

TABLE V

SRC-II FUEL OIL TEST PROGRAM

OBJECTIVE:
Assess operation and emissions using SRC-II Fuel Oil in a utility size boiler.

PHASE I - INITIAL BASELINE TESTING
- o Develop NO_x reduction trends by staged combustion techniques
- o 29 Emissions test (24 full load/5 half load)

PHASE II - SRC-II FUEL OIL TESTING
- o Characterize nominal operation emissions levels and performance
- o Establish acceptable minimum NO_x levels (starting with trends of Phase I) and characterize emissions and performance at these conditions
- o 17 Emissions test (9 full load/6half load/2@3/4 load)

PHASE III - FINAL BASELINE TESTING
- o Operate boiler with No. 6 oil in same configurations as operating in Phase II
- o Characterize emissions and performance
- o 28 Emissions tests (13 full load/13 half load/2@ 3/4 load)

TABLE VI

LARGE SCALE SRC-II FUEL OIL BURN TEST AT CON ED

	EPA REQUIREMENTS	TEST BURN RESULTS
NO_X	400 PPM	175-300 PPM
SULFUR	85% REMOVAL	95% REMOVED
PARTICULATES	.03	<.03 (NO PRECIPITATOR)
HYDROCARBONS	-	< 3 PPM
CO	-	<50 PPM
SO_3	-	< 1 PPM
BOILER EFFICIENCY	-	COMPARABLE TO PETROLEUM FUEL OIL

COAL LIQUIDS (SRC-II) ARE IN MOST RESPECTS SUPERIOR TO RESIDUAL FUELS. THEY ARE MORE LIKE NO.2 DISTILLATES AND CAN SUBSTITUTE FOR PETROLEUM FUEL OILS IN THE MORE RESTRICTIVE ENVIRONMENTS.

PHASE III - Final Baseline Testing

The third phase testing involved measuring the combustion performance and emission levels while using the low-sulfur No. 6 fuel oil, with the boiler operating as close as possible to the operating conditions used during Phase II.

The Consolidated Edison test results, as shown in Table VI, indicated complete suitability of SRC-II coal liquids as a high quality boiler fuel. No operational problems were encountered and no deposits were observed. Combustion efficiency was comparable to that for the low-sulfur No. 6 fuel oil, as were the levels of carbon monoxide and hydrocarbon emissions. Modifications to burner equipment required to handle the SRC-II fuel oil are considered to be no more extensive than those required for similar variations in petroleum fuels. Particulate emissions for the SRC-II fuel oil were generally lower than for the No. 6 fuel oil, and were in all cases below the new source performance standards proposed by EPA (0.03 lbs/MM Btu).

All tests were run with no smoking and less than .03 lbs/MM Btu total particulates. While the higher nitrogen content SRC-II fuel oil produced higher NO_x emission levels than the low-sulfur No. 6 fuel oil, the difference was substantially less than the relative nitrogen contents of the two fuels. For example, the SRC-II fuel oil produced only 70% more NO_x than No. 6 fuel oil, even though its nitrogen content was more than four times as high. Furthermore, the tests showed that NO_x formation could be reduced substantially for both fuels (on the order of 35%) by staged combustion.

Based on the overall test results, it would be expected that a boiler currently capable of meeting the EPA requirements of 0.3 pounds per MM Btu for petroleum fuels will be capable of satisfying the proposed standard for coal-derived liquids (0.5 lbs/MM Btu - equivalent to 400 ppm) using the SRC-II fuel oil.

Product Applications Testing

The low viscosity and pour point characteristics of the SRC-II distillates are also attractive in industrial boiler and industrial cogeneration applications substituting for No. 2 fuel oil or natural gas. Demonstration burn programs in industrial boilers are being planned.

Use of SRC-II distillates in stationary gas combustion turbines is also of significant interest. The low levels of trace metals and inorganics suggest minimal difficulty in regard to turbine blade erosion or corrosion. The higher radiant heat effect on the combustor walls caused by the lower hydrogen content of the SRC-II distillate requires appropriate but not unique design. As with the boiler application, higher nitrogen content will require NO_x minimization measures. Several DOE and EPRI developmental programs for NO_x control of coal liquids in combustion turbines are underway or planned.

The widespread use of combustion turbines in industry and by electric utilities, as well as the generating efficiency improvement offered by new combined cycle plants in conjunction with combustion turbines, represents an attractive market opportunity for SRC-II distillate coal liquids.

The medium-speed diesel (railroad locomotive, marine engines) appears to be another potential application for SRC-II coal liquids to displace petroleum fuels. Other applications being studied by potential users include the automotive turbine, reheat furnace fuel in the steel industry and reformer feedstock for fuel cells. All in all, the products to be derived from coal liquefaction processes like SRC-II can, over time, displace a portion of our requirements for imported petroleum in a variety of end uses.

Summary and Conclusions

Large-scale demonstration of the SRC-II process is currently being pursued as the next step in establishing the capability for the conversion of our high-sulfur coal reserves into a spectrum of hydrocarbon products to displace imported petroleum.

Under contract to the Department of Energy, Gulf is completing a preliminary analysis of the design, environmental effects, market opportunities, related economics and site requirements of the demonstration plant and subsequent commercial plants. The demonstration program will involve engineering development and testing of the large-scale equipment necessary for coal liquefaction.

Large-scale testing in a utility boiler of SRC-II coal distillates from the Ft. Lewis pilot plant indicates complete acceptability in combustion performance and emissions. Testing and development for other applications of SRC-II produced coal liquids is planned, including combustion turbines and medium speed diesels.

"Literature Cited"

1 Schmid, B. K.; Jackson, D. M. "The SRC-II Process" presented at the Third Annual Internation Conference on Coal Gasification and Liquefaction, University of Pittsburgh, Pittsburgh, Pennsylvania, August 3-5,1976.

2 Schmid, B. K.; Jackson, D. M. "Recycle SRC Processing for Liquid and Solid Fuels", presented at the Fourth Annual International Conference on Coal Gasification, Liquefaction & Conversion to Electricity, University of Pittsburgh, Pittsburgh, Pennsylvania, August 2-4, 1977.

3 Jackson, D. M.; Schmid, B. K. "Commercial Scale Development of the SRC-II Process", presenthd at the Fifth Annual International Conference on Commercialization of Coal Gasification, Liquefaction and Conversion to Elec-ricity, University of Pittsburgh, Pittsburgh, Pennsylvania, August 1-3, 1978.

4 "Combustion Demonstration of SRC-II Fuel" Electric Power Research Institute (report to be published, Spring, 1979).

RECEIVED August 13, 1979.

Exxon Donor Solvent, Coal Liquefaction Process Development

W. R. EPPERLY and J. W. TAUNTON

Exxon Research and Engineering Company, P.O. Box 101, Florham Park, NJ 07932

This paper describes the status of the development of the Exxon Donor Solvent (or EDS) coal liquefaction process. It includes an overview of the jointly funded project and a brief description of the EDS process. It also includes a discussion of the project status, including a description of coal feed flexibility, hydrogen and fuel gas production alternatives and the progress in the construction of the 250 T/D pilot plant. Other communications have covered the R&D program, the outlook for commercialization, and the organization of the EDS Project(1,2,3,4,5,6,7).

The goal of the EDS coal liquefaction project is to develop the process to a state of commercial readiness. This means that the technology should be available at the end of the project to design and build a full-scale, pioneer commercial plant with a reasonable and acceptable level of risk.

In the EDS process development, bench scale research, operation of small pilot units, and engineering design and technology studies are being integrated with operation of a 250 T/D pilot plant and a 70 T/D FLEXICOKING* prototype unit. (*Service Mark)

EDS PROCESS DESCRIPTION

The process sequence shown in Figure 1 is designed to maximize liquid products. The feed coal is crushed, dried, and slurried with hydrogenated recycle solvent (the donor solvent) and fed to the liquefaction reactor in admixture with gaseous hydrogen. The reactor design is relatively simple: an upward plug flow design with operating conditions of 800-900°F and about 2000 PSI total pressure. The reactor effluent is separated by a series of conventional distillation steps into a recycle solvent depleted of its donor hydrogen, light hydrocarbon gases, C_4-1000°F distillate, and a heavy vacuum bottoms stream containing 1000°F+liquids, un-

0-8412-0516-7/79/47-110-071$05.00/0

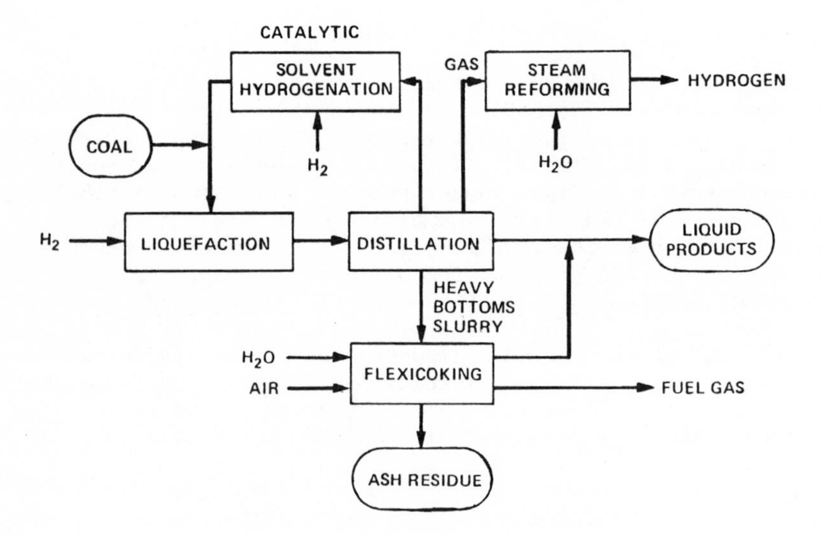

Figure 1. Exxon donor solvent process

converted coal, and coal mineral matter. The recycle solvent is hydrogenated in a conventional fixed bed catalytic reactor employing "off-the-shelf" hydrotreating catalysts.

The heavy vacuum bottoms stream is fed to a FLEXICOKING unit along with air and steam to produce additional distilled liquid products and a low BTU fuel gas for process furnaces. FLEXICOKING is a commercial petroleum process that employs an integrated coking/gasification sequence in circulating fluidized beds. This process is operated at low pressures (∿50 psi) and intermediate temperatures (900-1200°F in the coker and 1500-1800°F in the gasifier). Essentially all organic material in the vacuum bottoms fed to FLEXICOKING is recovered as liquid product or combustible gases. Residual carbon is rejected with the ash from the gasifier fluidized bed.

Process hydrogen is produced by steam reforming light hydrocarbon gases. An alternative method for hydrogen production is partial oxidation of the heavy vacuum bottoms stream or of coal.

The total liquid product is a blend of streams from liquefaction and FLEXICOKING. Product utilization studies indicate that the 350°F-fraction should be used in gasoline/petrochemical manufacture and the 350°F+ fraction in fuel oil applications. The latter fraction from Illinois #6 coal contains about 0.6 wt% sulfur and about 0.8 wt% nitrogen. These levels can be reduced further by subsequent treating, if needed, to meet emissions standards. The 350-650°F fraction may also be attractive as a turbine fuel (8,9).

EDS PROJECT STATUS

Exxon Research and Engineering Company (ER&E) has been engaged in coal liquefaction research since 1966. The jointly funded research and development project started in 1976 and is approaching the halfway point, entering the fourth year of a 6-½ year program. The U.S. Department of Energy (DOE) is providing 50% of the funding and the remaining 50% is being provided by The Carter Oil Company (an Exxon affiliate), Electric Power Research Institute (EPRI), Japan Coal Liquefaction Development Company (JCLD), Phillips Petroleum Company, and Atlantic Richfield Company. Atlantic Richfield Company and Japan Coal Liquefaction Development Company became sponsors in 1978, and additional participants are expected.

The project is functioning well under the provisions of the Cooperative Agreement with the Department of Energy, as reported elsewhere (7).

A broad technical program on the liquefaction portion of the process is being advanced essentially on schedule. The development of the vacuum bottoms processing is being expanded to include operation of a 70 T/D FLEXICOKING prototype at Baytown, Texas using the vacuum bottoms produced from the 250 T/D liquefaction pilot plant. In addition, new leads and understanding resulting from both project studies and relevant Exxon privately sponsored research work are being incorporated in the program.

Experience indicates that an important part of a normal process development is definition of solutions to operability and reliability problems that have been identified. The EDS process development is no exception. Potential mechanical problems associated with feed slurry preheat, slurry pumping, high pressure letdown valves and vacuum bottoms pumping have been identified and will be addressed in the 250 T/D pilot plant program. In addition, several process problems associated with the variety of coals processed have been identified and solutions defined. The status of both pilot plant construction and definition of solutions to process problems is presented in this paper.

250 T/D PILOT PLANT PROGRESS

Under the direction of Carter Oil, the construction of the 250 T/D pilot plant is proceeding on schedule. Figure 2 shows an artist's conception of the completed plant which details the relative position of the administration building, the coal preparation facilities, the process area and the product tankage areas. The cost outlook for the completed plant is 101 M$ compared to the initial estimate of 110 M$.

The schedule leading to mechanical completion in the fourth quarter of 1979 is shown in Figure 3. The control house is scheduled for completion in May. Final electrical facilities including substations and interplant lines are scheduled for completion in June. Coal preparation facilities and operational tankage are to be commissioned in August followed by completion of the solvent hydrogenation section in October. Final mechanical completion will be accomplished with turnover of the liquefaction reactor and fractionation sections, and is projected for mid November.

Startup activities are scheduled to begin this summer as sections of the plant are completed. Initial shakedown will involve startup solvent preparation and oil circulation throughout the plant. Coal-in operations are expected by January, 1980.

The operations of the 250 T/D pilot plant are designed to demonstrate the operability of the EDS liquefaction section and obtain the scaleup data required for design of a commercial facility. Key objectives are demonstration of unit operability,

Figure 2. Artist's rendering of the 250-ton/day pilot plant

design data acquisition and product yield and quality confirmation. Demonstration of sustained operation at suitable solvent quality with satisfactory operation of pumps, letdown valves and block valves in slurry service is also a key objective. Scaleup data from the slurry preheat furnaces, liquefaction reactors, slurry drier and vacuum fractionation unit when combined with other studies will provide the necessary input for commercial facility design. Data on product yields and qualities will be correlated with data from the 50 lb/D and 1 T/D pilot plants, and the products generated are to be used in product tests aimed at defining potential issues with regard to commercial utilization.

FEED COAL FLEXIBILITY

Efforts in 1978 verified that the EDS process can be applied to a wide variety of coal types including bituminous, subbituminous coals and lignites are more difficult to process, and this aspect will be discussed subsequently.

Illinois #6 bituminous coal and Wyoming subbituminous coal were specified initially as project coals and DOE, Carter Oil, EPRI and JCLD chose additional coals for evaluation. Operations of the 250 T/D liquefaction pilot plant are to include processing three coals; Illinois #6, a subbituminous coal, and a third coal to be selected by the project sponsors.

Analyses of coals which have been processed in the continuously operated pilot plants are listed in Table 1. Process liquid yields from the liquefaction step for these coals are shown in Figure 4 for different residence times in the liquefaction reactor. Longer residence time increases conversion of coal to liquids, but also increases hydrocracking of liquids to gas. As a result, there is an optimum time for each coal for maximizing liquid yield. An approximate explanation of these yields based on the coal analyses shown in Table 1 can be made. Higher yields correlate with high volatile matter, sulfur content, and reactive fractions and are, of course, inversely proportional to ash content.

The overall process yields which include liquids from both liquefaction and FLEXICOKING are shown in Figure 5 at the preferred conditions for project coals. The bituminous coals give total liquid yields in the 43-45% range, the subbituminous coal produces about 40% liquid, and the lignites produce 33-53% liquid. Yields have potential for being increased using process improvements currently under investigation. It should be noted that the liquids recovered from FLEXICOKING for the Burning Star coal have been higher than for any other coal studied.

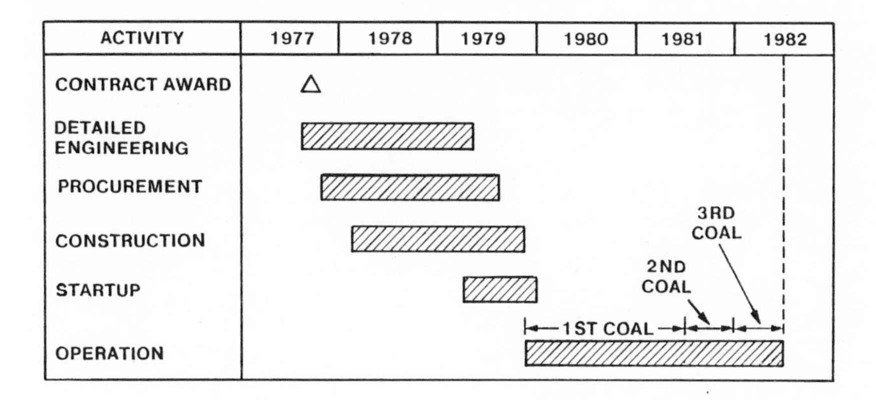

Figure 3. EDS 250-ton/day pilot plant schedule

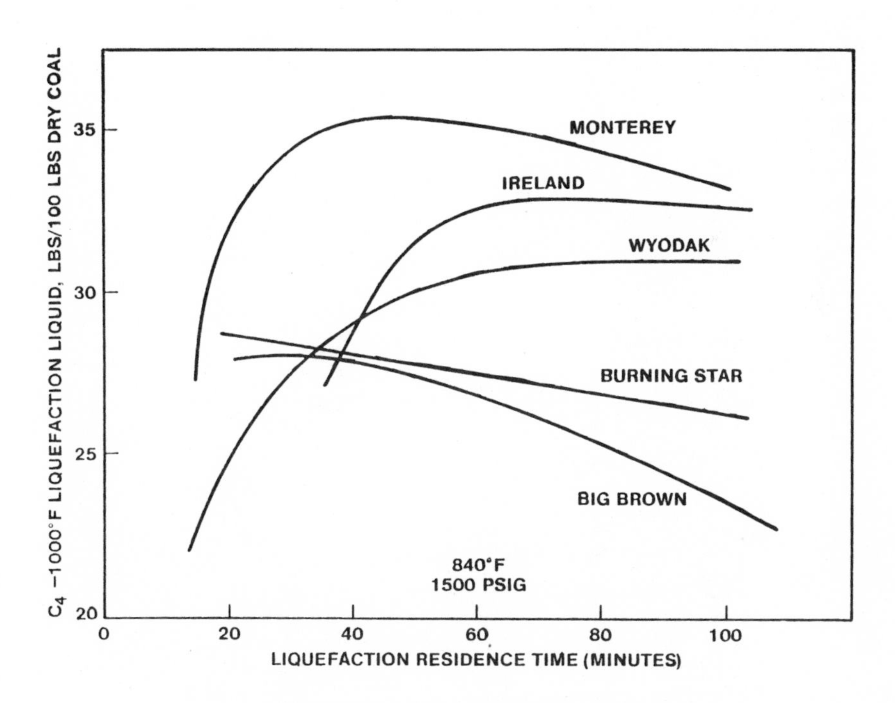

Figure 4. Liquid yield response to liquefaction residence time

TABLE 1

COAL ANALYSES

Coal Mine	Illinois #6 Monterey #1 (Bituminous)	Illinois #6 Burning Star #2 (Bituminous)	Pittsburgh Seam Ireland (Bituminous)	Wyoming Wyodak (Sub-bituminous)	Texas Lignite Big Brown
Elemental Analysis, Wt% Dry Coal					
Carbon	70.1	69.9	74.0	68.5	62.0
Hydrogen	5.1	4.9	5.2	4.9	4.8
Oxygen by Difference	10.6	10.4	6.3	17.2	14.5
Sulfur					
Total	4.1	3.1	4.3	0.5	1.2
Pyritic	1.3	-	2.0	0.1	0.3
Sulfate	0.1	-	0.1	0.1	0.1
Organic	2.7	-	2.2	0.3	0.8
Nitrogen	1.2	1.2	1.2	1.1	1.1
Chlorine	0.1	-	0.1	0.02	0.02
Ash	8.9	10.5	9.0	7.8	16.4
Ash (SO_3 Free)	8.8	10.0	8.8	6.6	14.2
Proximate Analysis, Wt% Dry Coal					
Fixed Carbon	47.3	50.8	51.9	46.3	39.2
Volatile Matter	41.8	38.7	39.1	44.5	44.5
H/C Atmoic Ratio, Dry Coal Basis	0.87	0.84	0.84	0.86	0.92
Petrographic Composition, Wt%					
Reactive	79.6	78.2	73.6	80.7	65.3
Unreactive	8.4	11.3	14.3	9.0	16.3
Mineral Matter	12.0	10.5	12.1	10.3	18.4
Vitrinite Reflectance, % Max.	0.45	0.52	0.65	0.38	<0.4

Inspection data for the products (from liquefaction and FLEXICOKING) from three coals are presented in Table 2. The product inspections indicate higher levels of nitrogen than found in similar fractions of petroleum. The sulfur levels in the products reflect the sulfur levels of the coals and are consistent with the anlayses presented in Table 1. Studies have shown that lower nitrogen and sulfur levels can be achieved should environmental standards require the lower values. (8,9)

Operations of the pilot plants on the various coals have indicated that the younger subbituminous coals and lignites are more difficult to process than are the bituminous coals. This is believed to be primarily due to the higher oxygen and/or organically associated calcium content of younger coals.

The difficulty of processing younger coals is illustrated in Figure 6 in which the viscosities of coal liquefaction bottoms derived from the various coals are shown. These viscosities are a direct measure of the ease of processing the various coals from a mechanical standpoint, e.g. pumping the bottoms from a vacuum fractionator or pumping the bottoms into a coking or gasification reactor. The approximate upper viscosity level for pumpability is shown as 50 poise. Figure 6 shows the viscosities of the bottoms from the younger coals to be higher than the viscosities of bottoms from bituminous coals. However, these higher viscosities can be reduced to pumpable levels with longer liquefaction reactor residence times under typical EDS conditions.

The high calcium content of the younger coals has led to the formation and deposition of calcium carbonate in the liquefaction reactor in the form of wall scale and oolites which were first observed in German operations (10). These deposits form as calcium salts of humic acids in the coal decompose under liquefaction conditions. The deposits continue to grow with time and could lead to unwanted solids accumulation in the reactor itself as well as fouling of downstream equipment (11). Data shown in Figure 7 indicate the accumulation rate of the calcium carbonate in the liquefaction reactor for different coals under typical EDS conditions as well as two methods for controlling the solids build-up.

One method of control is to use solids withdrawal from the liquefaction reactor coupled with strainers upstream of critical equipment such as valves, instruments, and pipe bends. In addition, reactor cleaning by chemical means during normal reactor turnarounds would be used to insure the required onstream time. An estimate of the calcium carbonate accumulation rate based on pilot unit experience is shown as the dashed line in Figure 7. This concept for calcium carbonate control is to be demonstrated in the 250 T/D pilot plant during operations on a subbituminous coal.

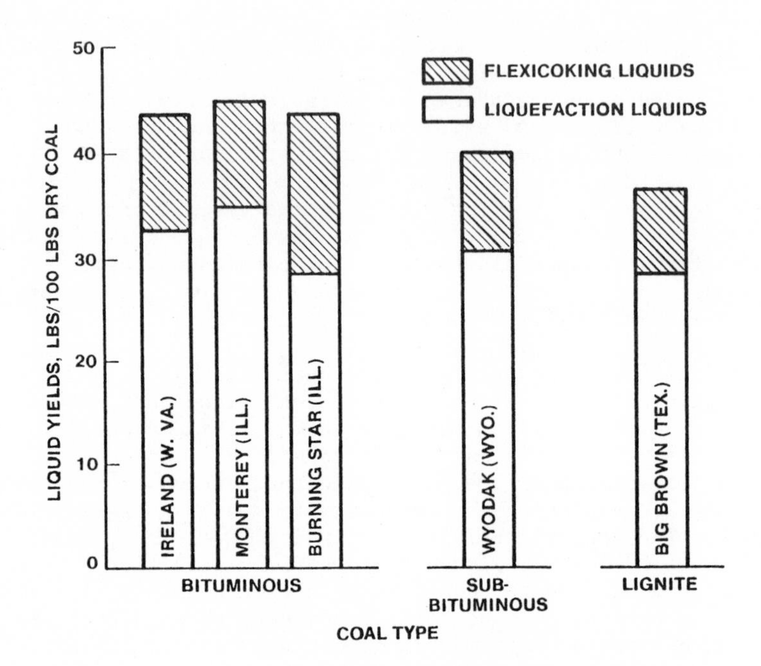

Figure 5. Preferred liquefaction/coking liquid yields in EDS process

TABLE 2

EDS PRODUCT INSPECTIONS

PRODUCT FRACTION	ILL. BITUMINOUS S, N WT%	WYO. SUBBITUMINOUS S, N WT%	TEX. LIGNITE S, N WT%
C_5–350°F	0.03, 0.07	0.003, 0.04	0.005, 0.06
350-650°F	0.01, 0.1	0.03, 0.3	0.1, 0.3
650°F+	1.0, 1.3	0.2, 1.2	0.3, 1.2
350°F+	0.6, 0.8	0.1, 0.8	0.2, 0.8

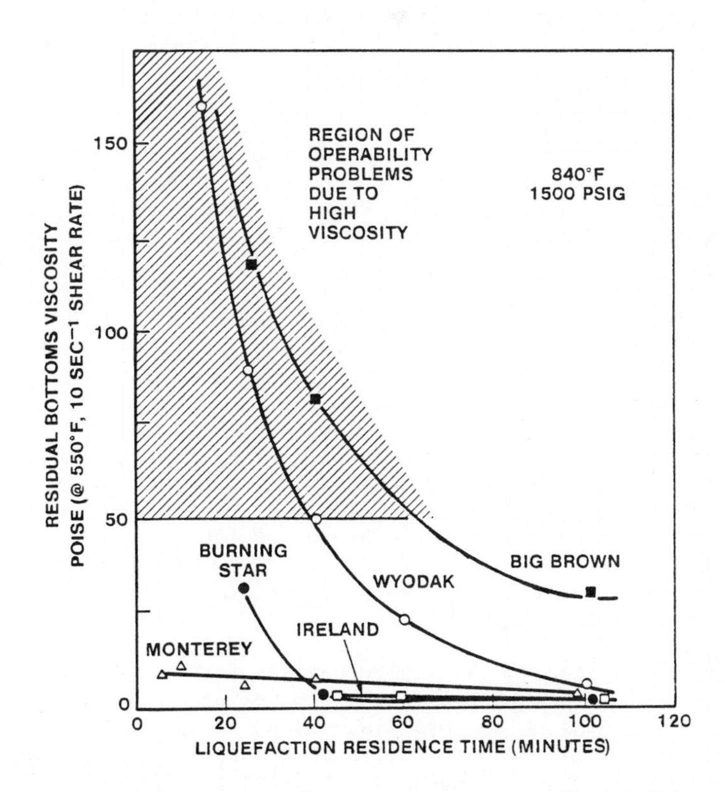

Figure 6. Residual liquefaction bottoms viscosity is an effective index to operability for different rank coals

Another method of calcium carbonate control is the use of pretreatment of coal with SO_2 to render the calcium innocuous as calcium sulfate. This technique was discovered in Exxon funded research and was subsequently made available to the project. The SO_2 reacts with the calcium in the coal and is then hydrolyzed to form the sulfate which does not form reactor deposits under EDS conditions. Inspections of reactors used to process SO_2 pretreated coal have indicated the presence of only insignificant amounts of the calcium carbonate.

The mechanical method of controlled scaling is the preferred method because of its simplicity and more favorable economics.

BOTTOMS PROCESSING

In the development of coal liquefaction processes considerable effort has been concentrated on the coal liquefaction part of the process. In contrast, less effort has been directed toward utilization of the coal liquefaction residue or vacuum tower bottoms.

Utilization of this stream, which contains one third to one half of the available carbon in the feed coal, is necessary to achieve hydrogen and plant fuel balances for the overall process, good carbon utilization and minimum cost. Alternatives for hydrogen and fuel production are depicted in Figure 8. The primary carbon sources for hydrogen and plant fuel are bottoms, coal, and light hydrocarbon gas. Bottoms can be processed in a FLEXICOKING unit to produce additional liquids and plant fuel, and in a partial oxidation unit to produce plant fuel or hydrogen. Coal is an alternate feed to partial oxidation. Light hydrocarbon gas can be steam reformed to make hydrogen or burned directly as plant fuel.

ER&E has studied these alternatives for the utilization of coal liquefaction bottoms in the production of hydrogen and fuel gas and in doing so has had discussions of partial oxidation with Texaco and Shell. These studies have identified a potentially attractive processing sequence utilizing FLEXICOKING to produce additional liquids and plant fuel, and partial oxidation to produce hydrogen.

Both FLEXICOKING and partial oxidation are commercial processes for petroleum residue (12,13). In addition, partial oxidation has been utilized to generate Synthesis gas with coal as a feed (14,15). Coal liquefaction bottoms have been processed in small pilot units in recent studies including Exxon's 2 B/D FLEXICOKING pilot plant (3) and Texaco's 12 T/D partial oxidation unit (16). Studies in Exxon's unit have included EDS bottoms from Illinois and Wyoming coals while SRC-I, SRC-II, H-Coal and

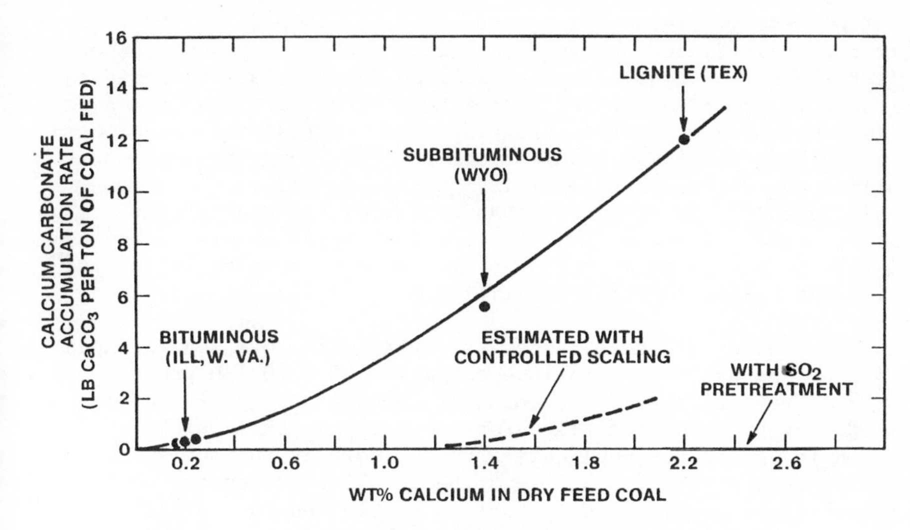

Figure 7. Calcium carbonate accumulation depends on coal source

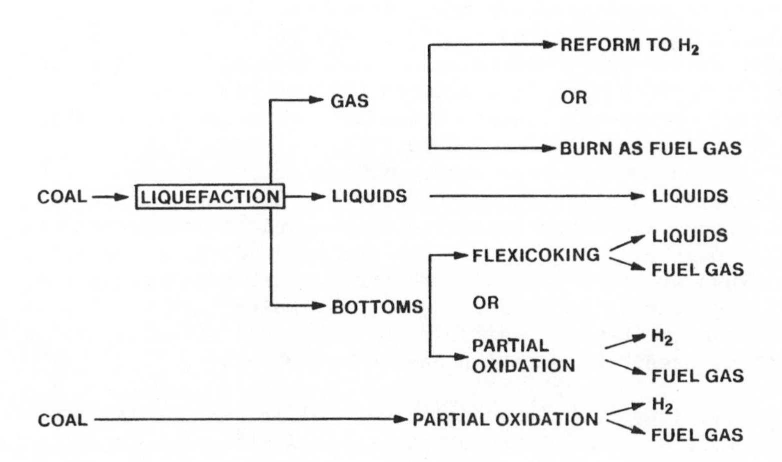

Figure 8. Fuel gas/hydrogen production alternatives

EDS bottoms are known to have been processed in Texaco's unit.

These studies have defined technical issues which require further study. These issues are noted in Table 3 and are derived from the differences between coal liquefaction bottoms and petroleum residue or coal. For FLEXICOKING, a principal issue is the impact of the high mineral matter content on particulate generation/control and gasifier slagging. For partial oxidation, one major concern is that high bottoms viscosity and thermal instability could limit applicability of the process to feeds containing more product liquid than is economically attractive. The alternative of feeding liquefaction bottoms as a solid would likely reduce overall process efficiency and require solidification and solids handling facilities.

Resolution of these issues for FLEXICOKING has led to expansion of the program to include operation of a 70 T/D prototype unit. The anticipated schedule for completion of this supplemental program for coal liquefaction bottoms FLEXICOKING is shown in Figure 9. Engineering design work is currently underway to modify the FLEXICOKING prototype to allow processing coal liquefaction bottoms. Construction is planned to start the first of next year with a February, 1981 mechanical completion. Operations are planned for eighteen months on bottoms from two coals generated by the 250 T/D liquefaction pilot plant.

ER&E discussions with Texaco and with Shell on bottoms processing are summarized herein. Texaco has indicated that its partial oxidation process could be applied to coal liquefaction bottoms on a commercial scale and that operation of their 12 T/D pilot plant with coal liquefaction bottoms representative of a projected commercial feedstock would be adequate to set the design basis for a commercial facility. Texaco indicated that three to four years after successful operation of the 12 T/D unit a commercial facility could be ready for startup. In initial discussions, Shell has indicated that development of the Shell/Koppers partial oxidation process for coal liquefaction bottoms would involve operations of both their 6 T/D pilot plant and their 150 T/D demonstration unit. It was estimated that the 150 T/D facility might become available in the late 1980/early 1981 time frame for possible operation on vacuum bottoms.

Discussions with Texaco and Shell will continue in order to pursue further application of partial oxidation for coal liquefaction bottoms.

EDS PROCESS IMPROVEMENTS

Inherent in the development of the EDS process is the belief that there are significant opportunities for process improvements.

TABLE 3

ISSUES IN COAL LIQUEFACTION BOTTOMS PROCESSING

COAL LIQUEFACTION BOTTOMS PROPERTIES	POTENTIAL PROCESS DEVELOPMENT ISSUES
• HIGH ASH/SOLIDS LEVEL	• GASIFIER SLAGGING • PARTICULATE GENERATION/CONTROL
• HIGH VISCOSITY/THERMAL INSTABILITY	• BOTTOMS PUMPABILITY • FEED CONTROL/DISTRIBUTION

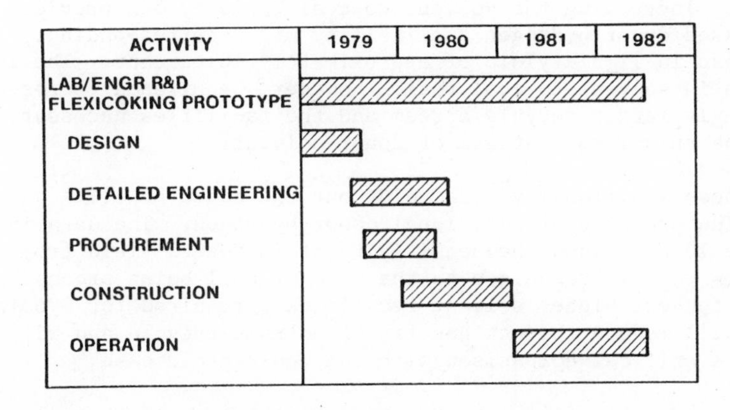

Figure 9. EDS coal liquefaction bottoms flexicoking development

This philosophy was incorporated into the participation agreements between ER&E and project sponsors. Potential process improvements are being brought into the project from Exxon's privately funded research, and are also being identified within the project. In addition, sponsors are suggesting improvements based on non-confidential information. Currently active process improvements are directed toward improving product yields, process operability and process efficiency.

Figure 10 shows one possible way of increasing liquid yields for certain coals. The data indicate that the yield of Illinois coal liquids (ex coking) from Illinois (Monterey) coal can be increased from 34-45 percent of dry feed coal by recycling coal liquefaction bottoms. This processing technique increases the residence time of the heavy bottoms in the liquefaction reactor and in this way increases liquid yield. As shown by Figure 5, longer residence time without recycle does not lead to the same increase because bottoms conversion to liquids is offset by hydrocracking of light liquids to gas. This can be seen by comparing the liquid yields in Figure 5 for the Monterey coal at 40 minutes residence time (coal conversion of 52%) and the liquid yield at 100 minutes residence time (coal conversion of 57%).

Figure 10 also shows initial data on the same processing technique applied to Wyoming coal. In this case insignificant yield increases were observed at the standard solvent-to-coal ratio. Increasing the solvent-to-coal ratio by 50% provides increased donor hydrogen availability and a corresponding increase in liquid yield of approximately 10 percent. The increase in solvent-to-coal ratio, however, requires a correspondingly larger recycle stream and the facilities necessary to process this larger stream of donor solvent.

These additional yields point out the added benefits gained from the presence of additional donor hydrogen. The data in Figure 10 also show the sensitivity of increased yield from the bottoms recycle technique to the type of coal being processed. The attendent higher solvent recycle rate required for Wyoming coal will reduce the net benefit of bottoms recycle and will require critical comparison with the non-recycle case.

In order to utilize higher yields, the overall thermal efficiency of an energy balanced plant must also increase. In Figure 11, the Illinois base case of 43 percent net liquids and a thermal efficiency of about 63 percent is depicted. This is based on the 1975-76 study design using the higher heating values of the total feed and products (3). Restricting attention to only energy balanced cases and the assumptions of the 1975-76 study design, a liquid yield of 50% could only be achieved by increasing thermal efficiency to about 70%.

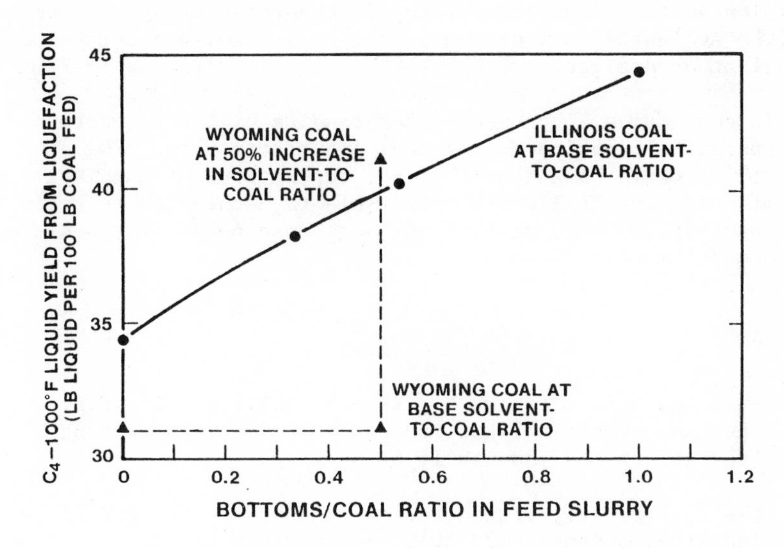

Figure 10. Simulated bottoms recycle provides increased liquefaction liquid yields

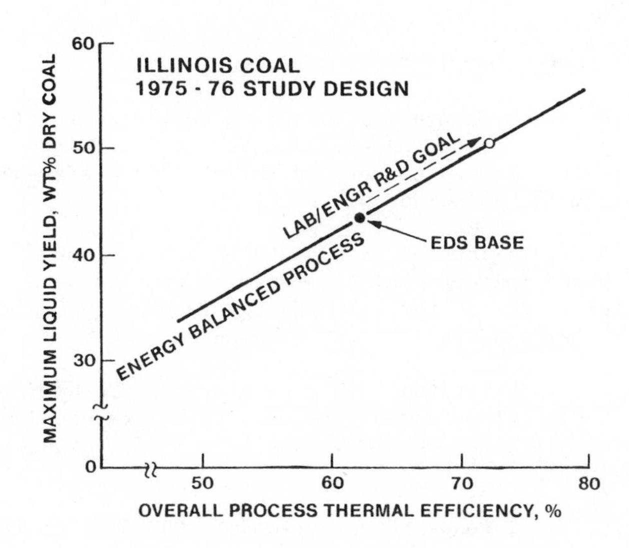

Figure 11. Liquid yield and efficiency related

Several processing schemes have been identified for increasing liquid yield. Ideas for achieving a higher thermal efficiency are being incorporated in the engineering program and will be evaluated as additional insight is achieved through additional study designs.

In conclusion, increased understanding of the requirements for successful development of coal liquefaction for a wide variety of coals has been achieved. Operations of the large liquefaction and FLEXICOKING pilot plants, scheduled to begin in 1980 and 1981, should provide the data base needed for scale up to commercial size.

"Literature Cited"

1 Furlong, L.E.; Effron, E.; Vernon, L.W.; Wilson, E.L. The Exxon Donor Solvent Process, Chemical Engineering Progress, p. 69, August 1976.

2 Swabb, L.E. Jr., Liquid Fuels From Coal: From R&D to an Industry, Science, 199, 619 February 10, 1978.

3 EDS Coal Liquefaction Project Technical Progress Reports prepared for the Department of Energy; Interim Report FE-2353-13, EDS Commercial Plant Study Design, February,1978 1976 Phase IIIA Annual Report FE-2353-9, December, 1977; Phase IIIA Final Report, March, 1978; 1977-1978 Phase IIIB Annual Report, FE-2893-17, September, 1978.

4 Epperly, W.R.; Taunton, J.W., Exxon Coal Liquefaction Process Development, paper presented at the Thirteenth Intersociety Energy Conversion Engineering Conference, August 20-25, 1978, San Diego, CA.

5 Epperly, W.R.; Taunton, J.W., Status and Outlook of the Exxon Donor Solvent Coal Liquefaction Process Development, paper presented at the Fifth Energy Technology Conference, February 27 - March 1, 1978, Washington, D.C.

6 Epperly, W.R.; Taunton, J.W., Development of the Exxon Donor Solvent Coal Liquefaction Process, paper presented at the Philadelphia AIChE Meeting, June 4-8, 1978, Philadelphia, Pennsylvania.

7 Epperly, W.R., Cooperative Agreement, EDS Coal Liquefaction Project, paper presented at the Regional Symposium of the National Contract Management Association, October 27, 1978, Houston, Texas.

8 Quinlan, C.W.; Siegmund, C.W., Combustion Properties of Coal Liquids from the Exxon Donor Solvent Process, paper presented at the ACS National Meeting, Anaheim, CA., March 14,1978

9 Fant, B.T.; Barton, W.J., Refining of Coal Liquids, presented at the API 43rd Mid-year Meeting, Toronto, Canada, May 8-11, 1978.

10 Wu, W.R.K.; Storch, H.H., Hydrogenation of Coal and Tar, U.S. Dept. of the Interior, Bureau of Mines, Bulletin 633 1968.

11 Given, P.H., et. al., Characterization of Mineral Matter in Coals and Coal Liquefaction Residues, EPRI Annual Report AF-832, Research Project 3361, Pennsylvania State University, December, 1978.

12 Stretzoff, S., Partial Oxidation for Syngas and Fuel, Hydrocarbon Processing, December, 1974, p. 79-87.

13 Blaser, D.E.; Edelman, A.M., FLEXICOKING for Improved Utilization of Hydrocarbon Resources, paper presented at the API 43rd Mid-year Meeting, Toronto, Canada, May 8-11, 1978.

14 van der Burgt, M.J.; Kraayveld, H.J., Technical and Economic Prospect of the Shell-Koppers Coal Gasification Process, paper presented at the ACS National Meeting, Anaheim, CA., March 16, 1978.

15 Reed, T.L., Preliminary Design Study for an Integrated Coal Gasification Combined-Cycle Power Plant, EPRI Report AF-880 Research Project 986-4, Southern California Edison, August, 1978.

16 Robin, A.M., Hydrogen Production from Coal Liquefaction Residues, EPRI Final Report AF-233, Research Project 714-1, Texaco, Inc., December, 1976.

RECEIVED July 2, 1979.

6

The H-Coal Process

C. D. HOERTZ and J. C. SWAN

Ashland Synthetic Fuels, Inc., P.O. Box 391, Ashland, KY 41101

The H-Coal process is a development of Hydrocarbon Research Inc. (HRI). It converts coal by catalytic hydrogenation to substitutes for petroleum ranging from a low sulfur fuel oil to an all distillate synthetic crude, the latter representing a potential source of raw material for the petrochemical industry. The process is a related application to HRI's H-Oil process which is used commercially for the desulfurization of residual oils from crude oil refining.

The H-Coal process has been thoroughly tested on bench scale and process development units. This work was initiated over 14 years ago and has continued until now through funding arrangements with government and industry. As a result, there is a data base of more than 60,000 hours at the bench scale level and 10,000 hours on a 3 TPD Process Development Unit. There is now a large scale pilot plant under construction that is designed to process 200 to 600 TPD of coal. This will be the last step necessary to establish technical and economic feasibility for H-Coal and provide design data for a commercial plant.

The H-Coal process is primarily a liquefaction system but does produce significant quantities of SNG and LPG. Figure 1 presents a schematic of the process. Briefly, coal is cleaned, dried, pulverized and slurried with process-derived oil in the preparation section. It is then pumped to reactor pressure, mixed with hydrogen, heated, and charged to the reactor. There, the coal, recycle oil and hydrogen react in the presence of a catalyst at pressures up to 3500 psig and temperatures to 850°F. Depending on the severity selected, the product slate can be an all distillate material or a liquefied residuum with only a small amount of distillate. After leaving the reactor, the liquid effluent is treated to provide a low-solids recycle oil which is used to slurry the coal. The balance of the liquid is fractionated into distillate products and ash-containing residuum. The heavy ends can be further treated to recover additional ash-free

0-8412-0516-7/79/47-110-091$05.00/0

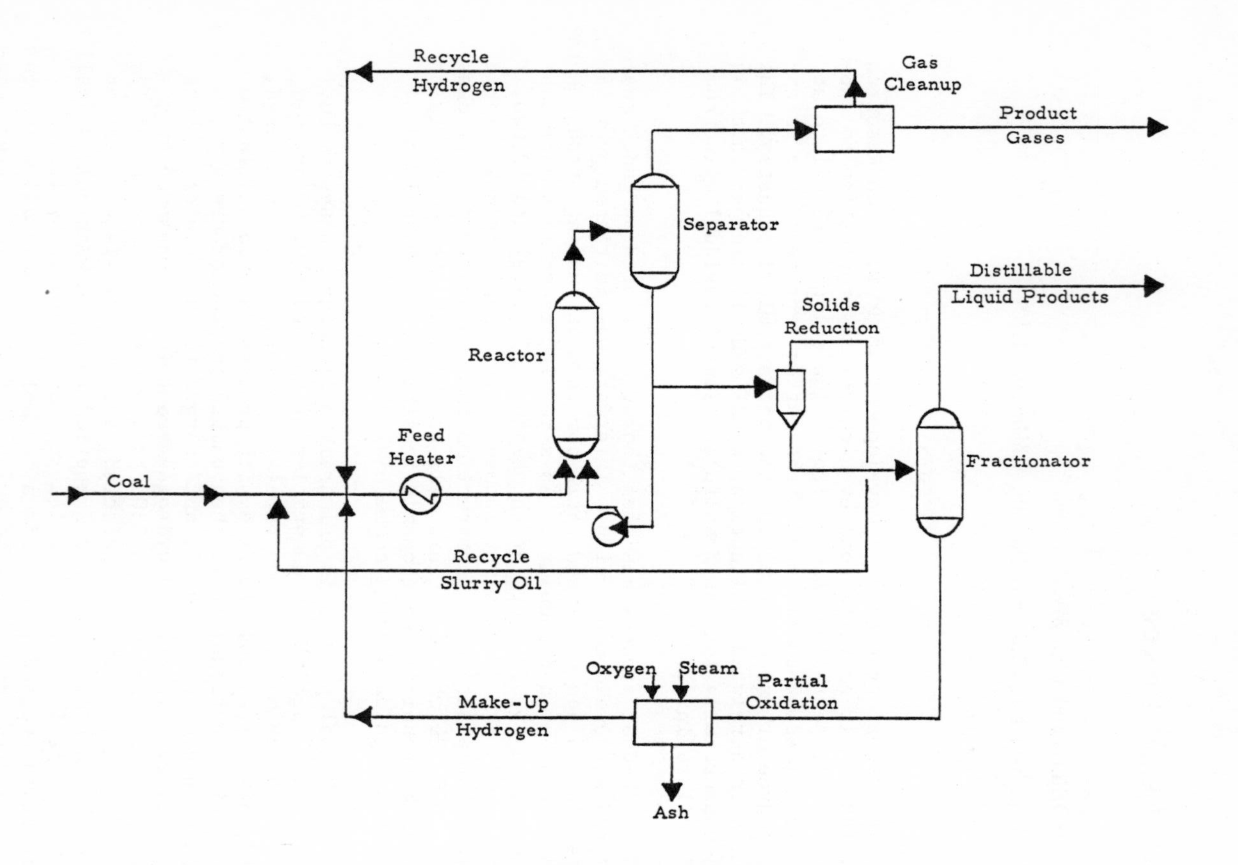

Figure 1. H-Coal process schematic

hydrocarbons or used as feed to a hydrogen plant. Figure 2 indicates the range the product slate can attain, depending upon commercial requirements.

The reactor is the key to the versatility of the H-Coal process. Figure 3 is a simplified diagram of the reactor. The concept involves a catalyst bed that is kept in an expanded or ebullated state by charging the feed and additional recycle oil to the bottom of the reactor. The products, including unreacted coal and ash, flow through the catalyst and are removed from the reactor at a point above the top of the catalyst bed. An external separator removes gaseous products and recycle hydrogen from the liquid.

Because the catalyst bed is constantly in motion, a portion of the catalyst can be routinely withdrawn and replaced with fresh catalyst. In practice, perhaps one percent of the reactor inventory would be replaced daily thus maintaining a high level of activity. In addition, this type of reactor will permit a high degree of isothermal operation and achieve a high level of efficiency through direct utilization of the energy generated by the reaction.

At the present time, a consortium of industry and government is funding an H-Coal pilot plant being constructed at Catlettsburg, Kentucky. Table I provides a summary of the project. The plant has been designed to process from 200 to 600 TPD of both bituminous and subbituminous coal, producing a nominal 600 to 1800 BPD of product. The cost is presently estimated to be $275 million including $35 million for research and engineering, $115 million for plant construction and $125 million for 2 years operation and subsequent dismantling. The funding group includes the Department of Energy, the State of Kentucky, the Electric Power Research Institute, Standard Oil of Indiana, Mobil, Conoco Coal Development and Ashland. Construction is approximately 70 percent complete with mechanical completion scheduled later this year.

The objectives of the pilot plant program are summarized in Table II. The plant is sized large enough to demonstrate mechanical operability of prototype and commercial equipment in the environment of coal conversion process conditions. At the same time, substantial quantities of products representative of commercial operations will be available for evaluation and development of downstream processing and markets. The fully integrated pilot plant will verify yield structure and supply design and other engineering data required for the design of a commercial plant. Finally, actual operation of the equipment over extended periods will allow extensive evaluation of materials of construction and development of maintenance requirements.

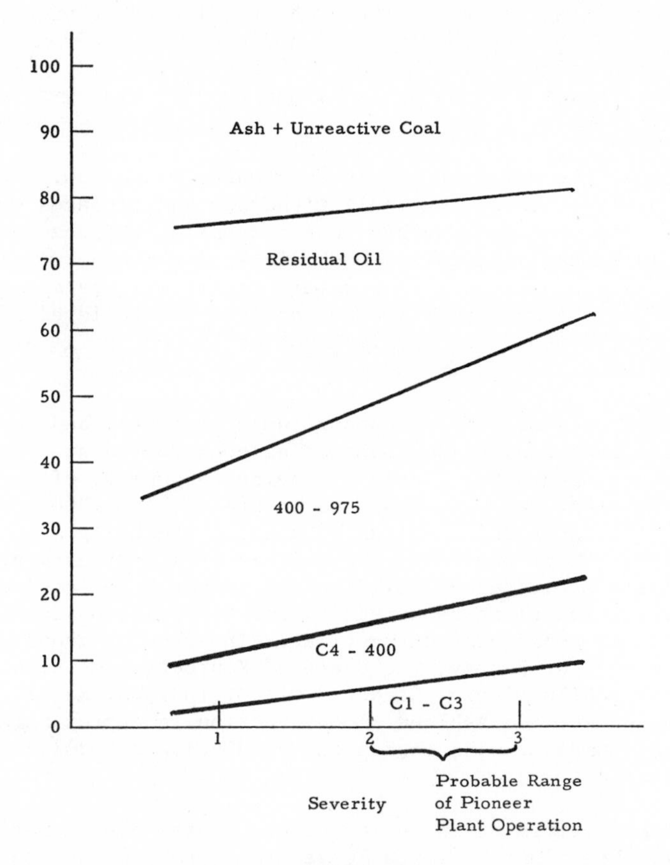

Figure 2. Yield vs. severity

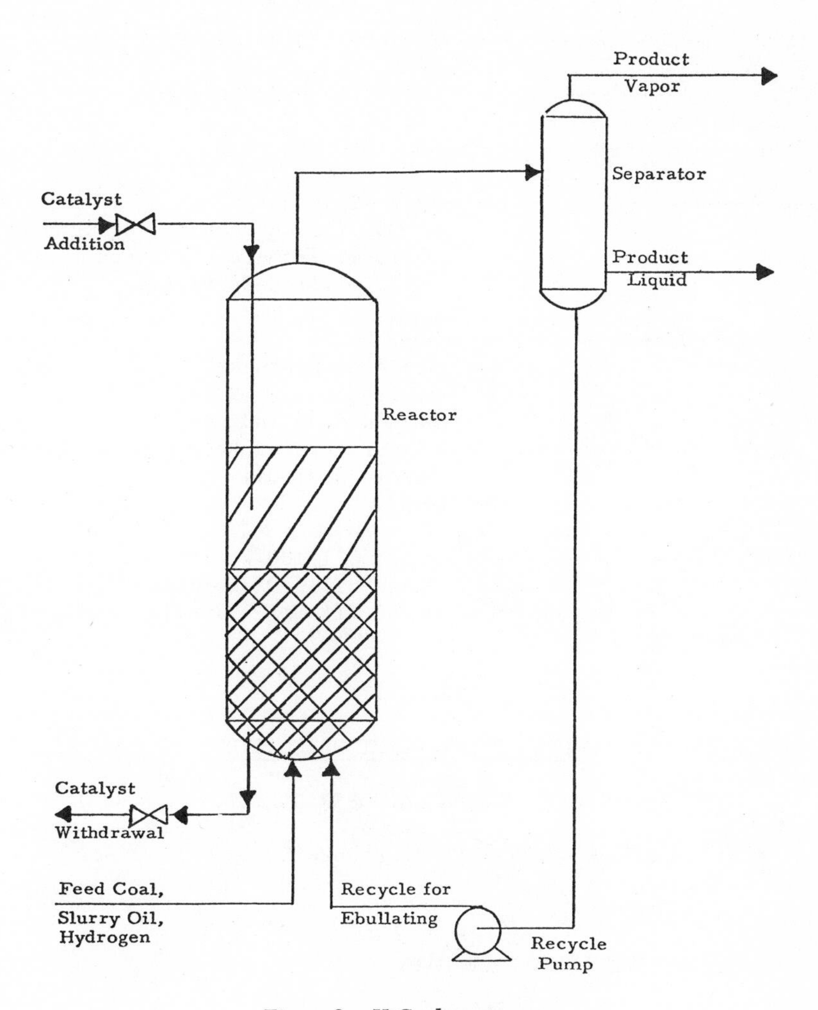

Figure 3. H-Coal reactor

TABLE I

H-COAL

PILOT PLANT FACT SHEET

OBJECTIVE	200-600 TPD Pilot Plant
TYPE	Catalytic Hydrogenation
YIELD	600-1800 BPD
COST	$275 million $ 35 million-Research & Engineering $115 million-Construction $125 million-Operation
FUNDING GROUP	DOE State of Kentucky Ashland Standard of Indiana Electric Power Research Institute Continental Oil Mobil
STATUS	Phase I-100% complete Phase II- 70% complete Phase III-Mid 1979-81

TABLE II

H-COAL PILOT PLANT OBJECTIVES

- DEMONSTRATE MECHANICAL OPERABILITY
- PROVIDE QUANTITIES OF PRODUCTS
- VERIFY YIELDS
- PROVIDE SCALE-UP DATA
- COLLECT ENGINEERING DATA
- DETERMINE MATERIALS OF CONSTRUCTION
- ESTABLISH MAINTENANCE REQUIREMENTS

As noted, the plant is expected to come on stream in mid-1979 after which a two-year operating program is scheduled. The plan calls for processing Illinois No. 6 or comparable bituminous coal in both the synthetic crude and fuel oil modes and then a subbituminous coal in the syncrude mode only. Each run is expected to be about three months in length to allow ample time for lineout and collection of engineering and operating data. Yields in the syncrude mode will show a high percentage of naphtha and light gas oil while in the fuel oil mode there will be over 80 percent of 400° plus distillate and residuum.

The plant is being constructed adjacent to Ashland's refinery at Catlettsburg, Kentucky. It occupies a 40 acre tract on the Big Sandy River. The Ashland refinery will supply a number of utility services including make-up hydrogen, thus reducing substantially the cost of construction.

Upon demonstration in the pilot plant that the H-Coal technology is commercially feasible, perhaps in early 1980, the tools will be in hand to proceed with commercial development. Presently activities are under way to procure a preliminary engineering design and capital and operating cost estimates for a full-size syncrude facility. Figure 4 is a simplified schematic of the plant as now conceived. The present design contemplates only the basic elements of a coal conversion plant except for a naphtha reformer and SNG separation facilities that are included primarily to recover hydrogen. All ash-containing residual material would be charged to a hydrogen plant.

The liquids output represents a combination of transportation and utility fuels as summarized in Table III. All of the naphtha is to be reformed on site to produce a very high aromatic stock. With an exceptional octane blending value, this stream will find ready application as a gasoline component, but perhaps more important, it is also a source of substantial quantities of petrochemical raw materials as noted in Table IV. The potential yield of BTX and phenolics along with the low boiling paraffins should make such a plant an important factor in the future supply picture for these materials.

The plant is sized to process 20,000 TPD of high sulfur bituminous coal and produce a nominal 50,000 BPD of liquid products including LPG and butanes. Approximately 30 million cubic feet of SNG would also be recovered along with minor amounts of sulfur and ammonia.

This would be a multi-train facility with each train being approximately the same size as the largest H-Oil plant presently in commercial operation. Also each train would have about 10 times the throughput as the pilot plant thus representing a

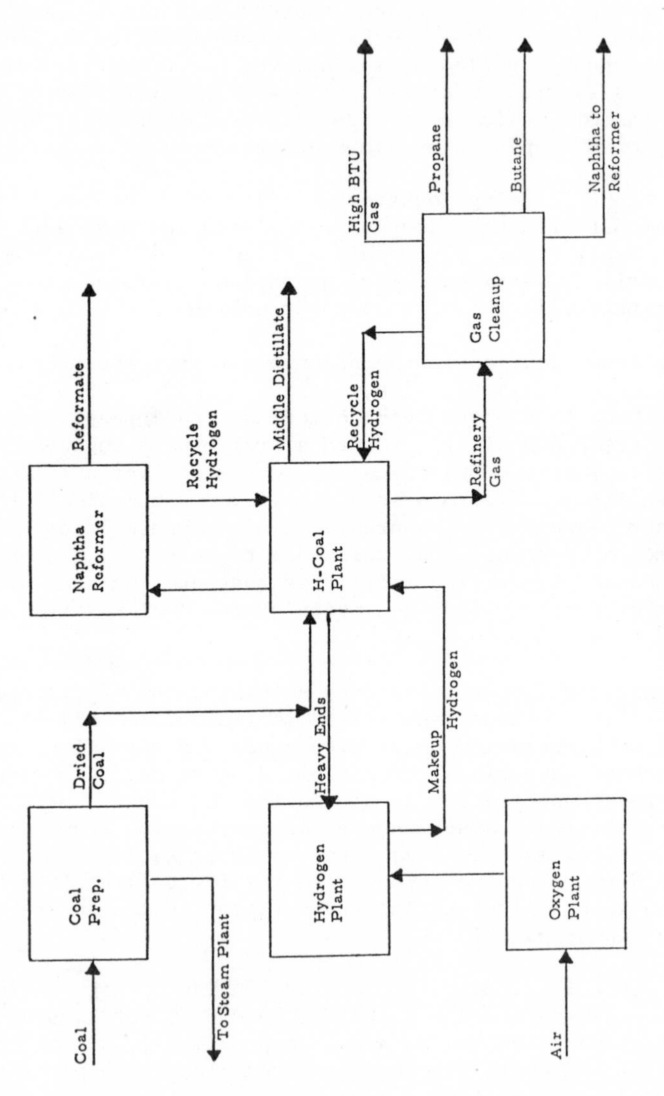

Figure 4. Commercial block flow diagram

TABLE III

COMMERCIAL PLANT

PRODUCT SLATE

Raw Coal	20,000 TPD
Products	
Reformate	15,300 BPD
Distillate (400-950°F)	27,900 "
Butane	3,300 "
Propane	3,500 "
Total	50,000 BPD
By Products	
Sulfur	570 LT/D
Ammonia	120 ST/D
SNG	31.7 MMSCFD

TABLE IV

PETROCHEMICAL POTENTIAL

Benzene	13.6 MM GPY
Toluene	24.5 MM GPY
Xylenes	32.2 MM GPY
Phenols & Cresols	170 MM lb/year
Butanes	222 MM lb/year
Propane	205 MM lb/year
Ethane	222 MM lb/year

reasonable scale-up for process equipment.

The proposed schedule for commercialization targets 1985 for start of production. This assumes that all environmental and permitting requirements will be met in time for construction to start in mid-1982. Based on past experience, this is an adequate time for preparation of an EIS and for PSD review but these activities are on the critical path so any additional time spent on unusual problems would certainly result in a day-for-day slippage in mechanical completion.

Best current projections for the proposed plant show a capital investment on the order of $1.0 billion in 1979 dollars. Economics based on this estimate will result in a rate of return far short of the hurdle rate required for total private investment. However, since a plant can not be on stream before 1985, the key question is the probability of coal liquefaction becoming commercially viable in the late 1980's. This is entirely possible under certain conditions such as the following:

- A differential developing in the escalation of future oil and coal prices. Many reliable sources predict rapidly rising oil prices after 1985 while coal will tend to follow a lower rate somewhat similar to general inflation. This divergence, depending upon its magnitude, will result in significant improvement in economics.

- Substantial differentials existing between fuel and petrochemical values of aromatics. In today's market, the recovery of petrochemicals from H-Coal naphtha would result in upgrading the average value of the product slate by more than $2/bbl. If this present differential is maintained or increases, it will be an important economic factor in the development of coal liquefaction.

- Financing. As is the case with any capital intensive technology, the economics of a coal liquefaction plant for a private investor can be affected to a large degree by the leverage of debt financing. Project financing in some form will be essential for many early investors, especially in the form of government guaranteed loans. This and other types of incentives will need to be established in order to foster the development of a versatile coal conversion industry.

All of these conditions are within the realm of possibility and give support to the feasibility of coal liquefaction in the mid-to-late 1980's.

The technology is rapidly approaching a state of development that can provide reliable commercial design data. Just as rapidly, the critical nature of the world energy outlook is becoming more definitized, making it obvious that any reasonable alternatives to crude oil as a source of fuel and petrochemicals must be evaluated for commercial potential as expeditiuosly as possible.

In cooperation with the government, programs are being established to move H-Coal and other technologies to the final stage of commercial development. Much work still remains, especially the resolution of economic risks, but there is a definite momentum building that can provide the necessary environment for a number of important projects to move forward in a timely manner.

RECEIVED August 1, 1979.

7

The Role of DOE's Energy Technology Centers

JAY R. BRILL[1]

Pittsburgh Energy Technology Center, 4800 Forbes Ave., Pittsburgh, PA 15213

Coal is a source of energy in relatively bountiful supply in our country and synthetic fuel applications of coal are a very vital topic for the American Chemical and Process Industry that you here represent. A greater use of this indigenous resource is a cornerstone of our National Energy Policy, and it is the very inventive genius of American enterprise that can turn this policy into a viable reality as the industry of our country has done so admirably in the past in stepping up to big issues that our nation has faced.

The word "dilemma" derives from two Greek words--"di" meaning two and "lemma" meaning assumption, and hence the definition presented earlier with a slight variance--"A choice between unpleasant alternatives." When one considers the alternative to petroleum supply uncertainties and shortfall, and certainly the events that we have seen around in recent days, the expanded use of coal and coal-derived synthetic fuels may not really be an unpleasant alternative, but indeed more one of a logical and necessary challenge.

The technology of coal and coal-derived synthetic fuels is a very sophisticated and complex business, every bit as sophisticated and complex as the high technology and synthesis of technologies that allowed our nation to so successfully conduct manned exploration of the moon. We face a tremendous challenge in expanding the use of coal in both direct combustion and in synthetic fuels, and doing this in an environmentally acceptable manner, which is one of our big challenges. This is a very major goal of the DOE Energy Technology Centers. I will relate the role of these Centers and where they fit into the strategy and activities of the DOE Fossil Energy Program. My focus will be on the Pittsburgh Energy Technology Center, the largest of a total of five that exist in the country. It's the one that has liquefaction as a major responsibility and the one with which I am associated.

[1] Current Address-Strategic Petroleum Reserves, DOE, Washington.

The Energy Technology Centers trace their genesis to the Bureau of Mines and became transferred assets from the Bureau of Mines when the Energy Research and Development Administration was formed in 1975, and then, in turn, when DOE was formed in late 1977. The Pittsburgh Energy Technology Center, or PETC, became an institution at our current location, about seventeen miles south of Pittsburgh, in 1948. But in going back and researching the archives, I found that its activities and basic mission in coal technology derive from the Bureau of Mines Pittsburgh Experimental Station established in 1919. For those of you who have been at 4800 Forbes Avenue, I think you can look at those buildings and see that they probably came around a lot before 1919.

Basically, our initiative in coal-derived synthetic fuels began at the end of World War II from the German technology and liquefaction experience, and this was the basic reason why the center at PETC was expanded from downtown Pittsburgh out to its current site. Then with the discovery of substantial oil assets in the mid-East War. So PETC's technology base on liquefied coal fuels has been in being since the end of World War II.

George Fumick, who is the Program Director for Fossil Energy in DOE, took a very important initiative in August of last year wherein he layed out a series of very specific lead laboratory responsibilities for all of the ETC's. I think this was very important. It happened just a little bit before I arrived on the scene at Pittsburgh, and it gave each center a specific series of purposes and responsibilities on which to focus their energy. I won't go over the centers out West that aren't directly related to the synthetic fuel business, but there is one in Laramie, Wyoming, one in Grand Forks, North Dakota, Bartlesville, Oklahoma, responsible for essentially enhanced oil recovery and internal combustion engines. But Morgantown in West Virginia and Pittsburgh have the basic lead responsibility in DOE for synthetic fuels. At Morgantown, they have a great responsibility for coal gasification, for fluidized-bed combustion, both atmospheric and pressurized, and gas-stream cleanup. We are the lead laboratory for coal liquefaction, coal process technology and a series of combustion activities that relate to magnethohydrodynamics combustion, direct combustion and coal/oil mixture; and I will talk a little bit about those later on.

Each of these centers has responsibilities in the DOE Fossil Energy Program and directly support the mission of that program. Each of the Center Directors reports directly to George Fumick who sits under the Assistant Secretary for Energy Technology, who is now John Deutsch.

Until recently, the product of the ETC's, for the most part has been applied research and technology. In the case of the

Pittsburgh ETC, this has been almost a wholly in-house program. Fossil energy is now in a process of decentralization of project management--a DOE-wide initiative to assign the management and execution of energy project to DOE field activities, not only to places like the Energy Technology Centers, but also to the various operations offices. The plan is currently in the process of implementation under the direction of Under Secretary Daly Myers with the full support of Secretary Schlesinger. This initiative will enhance the overall productivity of the Department by utilizing the technical and management resources of the field activities and by putting selective in-house focus on specific project problems in planning the in-house activities. One caution--I used the word selective. We still need to look to initiatives that should represent the cutting edge of technology. We must maintain that balance of technical project support--such as to the large liquefaction demonstration plants, for instance--and still keep a forward-looking technology base. We just must not mortgage away the future by putting all of our resources into paying today's bills.

Let me illuminate some of the key in-house technical activities at the Pittsburgh ETC, and then superimpose the near-term project management activities planned for transfer to Pittsburgh. Lastly, then, I would like to cover what I perceive as key initiatives to maximize our productivity. We are running into the same thing in this Department that I saw in the Department of Defense, and that is you are basically being asked to do more with less. In other words, we are drawing down on people and yet the workload, either real or apparent--I think it is real--is increasing. So there are three things you do. First, you can program continuing overtime, and that is certainly not the way to solve it. But I think the synergism of focusing our work to put our effort on major payoff items and depending more on contract effort--and we are going to be doing that--is really the way out of that particular dilemma we face there.

In our in-house activities, starting with combustion, we have three major areas. I think all of you are familiar with coal/oil mixture programs. This isn't a totally alternative source of energy but rather an "oil stretcher." Estimates for potential coal/oil mixtures use range from one-half to one-million barrels of oil displaced per day, or something like 2.50 to an upper limit of 5% for equivalent petroleum usage per day. We have a 700 hp. water-tube boiler burning coal/oil mixtures. With this new, highly instrumented facility, we are evaluating all aspects of coal/oil mixture combustion--flame characteristics, combustion efficiency, heat transfer, corrosion, erosion, pollutant emissions, and bottom ash removal. As I look at our total produce line and look for our near-term initiative, it seems that for the particular strategies where a coal/oil mixture can be used, it certainly has the greatest potential for being the earliest commercialization of any of the uses for coal, other than direct combustion.

A second area in our combustion work is in direct-combustion. Large industrial combustors are too costly to use for experimentation. However, we have a unique 500 lb/hr. pulverized coal/oil furnace which closely simulates the performance, in other words, the value of the heat per unit volume of commercial unit. Our main thrust in using this is that of resolving applied problems. With this combustor, we've studied the handling, pulverizing, combustion, and fouling characteristics of SRC-I fuel and operated it on these fuels first during October, 1974.

Our third area of combustion research involved magnetohydrodynamic or MHD power generation and the combustor development of such a system. Our work in MHD combustion is directed toward a part of a larger program which will result in an entire MHD system being operated in Butte, Montana sometime in Fiscal 1982. To support our MHD work, we have a one-megawatt atmospheric pressure combustor and a five-megawatt pressurized (6 atmospheres) combustor test facility.

MHD has the very interesting potential of producing electricity directly from coal more efficiently than present-day electricity generating power plants which now use coal, and I have seen numbers there of 50-55% versus around 35%, and doing this within acceptable standards for NO_x and sulfur effluent control.

A very key area of our in-house work is in liquefaction. Historically, we initiated the liquefaction program in the late 40's in Fischer-Tropsch work and coal hydrogenation. At that time, much of World War II German technologies had been tested and a further research program was built on top of that at PETC. A sizeable amount of work expended in Fischer-Tropsch in both catalyst research as well as pilot plant studies and design. Much of the design work in the SASOL I indirect liquefaction plant in South Africa and the pilot plant in Louisiana, Missouri, was obtained from the bank of information that was generated at PETC.

We have our work divided into process engineering, process chemistry, catalysis, and support technology. As an example, one of the indirect liquefaction projects, tube wall reactor, deals with the design and operation of high thermal efficiency catalytic reactors for syn-gas conversion. Other activities are coal liquefaction properties of coal minerals, the role of catalysts, coal liquid product stability, and environmental impact--to name a few.

Third-generation gasification research is going on at PETC.

The concept is a dilute-phase hydrogasification process in which coal is directly reacted with hydrogen to produce maximum yield of methane in the reactor. We are not, as an organization, competitive with industry either in hardware or in process work. Our objective really is to support and facilitate the industry. In this regard, we are working with Rocketdyne, who has been working on a similar concept with their unique reactor concept. That basically, is a spinoff from the space program. They are taking a design, as I understand it, from the F-1 million-pound liquid oxygen/kerosene rocket engine, and adapting that as a very fast reactor concept.

I would emphasize that environmental impact anlysis, the development of environmental control strategies, and energy conservation are an integral part of each project. In fact, we have a division that not only assures compliance with the various statutes in PETC's daily operations but also performs research on process specific, site specific environmental and energy conservation activities.

Let me now tie this in with our near-term project management activities.

Our role relative to project management has two dimensions: first, there are several projects in liquefaction and combustion that we will directly manage and these will be coming from Washington to the field; and secondly, we will be providing major technical support to the DOE project management offices for the large liquefaction demonstration plants. Let me expand on this a little bit.

We will be managing the 20 MW MHD combustor competitive prototype projects that are currently under way. One of these three contractors will be selected to develop and produce the combustor for a 50 MW MHD component development and integration facility scheduled for operation in Montana in FY 82. We will also be managing two coal/oil mixture demonstration projects. One will involve a utility steam generator and one is a blast-furnace operation.

In liquefaction we will have basically a series of industrial projects, industrial technology projects, such as R&D, process support, product upgrading, and a large number of relatively small dollar volume but important, and university projects to be transferred and aligning them into coherent work packages for transfer to field.

We are re-structuring the PETC organization from that of a traditional functional line division organization reflective of a total in-house research center to a matrix organization that will be responsive to our new role. A part of this organizational plan will include an Assistant Director for Project Management and his responsibility will reflect both direct project management functions and technical support functions focused on the large demonstration plants. We envision PETC technical support managers for the major plants who also will be a part of the overall Oak Ridge Operations Project Management Office and who will have the authority to call upon support from our in-house technical divisions in a matrix support role.

As my last thought, let me share some of our initiatives with you that I perceive can sharpen our focus and enhance our productivity.

We are becoming more customer-oriented to you here--our customers. I used to be in a business where there was a product line, such as an airplane, a missile or something like this. But our business is clearly to support you people who are really the customers of our business. Our task is to provide that bridging technology and support required for the private sector to adapt and move to commercialization. We need both more management level and technology working-level transfusion and dialogue with you people in energy industries. We are increasing industry coordination meetings, and I have initiated visits to the R&D organizations of the major companies. I encourage even more visits from the industry to PETC.

Our emphasis is on goal- and performance-oriented, not effort-oriented, activities and initiatives. In other words, if we are going to do something for a given year, it's not just to take so much money and work that for that year on an effort-oriented basis, but to strive toward certain specific goals or performances. We are maintaining a dynamic system of evaluating actual against planned progress. We need to push success and assess unsuccess. I find that the gold-watch syndrome kind of prevails: once things get started, they never stop. But I think it is important to recognize that every activity become successful and we need room for emerging technologies. So we are looking carefully at items which, after a few months or a year, did not achieve what we thought they should and to be cancelled to make room within the budget and within the resources for more exciting opportunities.

Systems analysis and systems synthesis are powerful tools to illuminate both technology gaps and exciting payoff activities where technology enhancement has substantial potential for economic payoff. We are evaluating the total system and avoiding suboptimization.

We need a stronger capability at PETC in determining

economic earned value and cost-benefit analysis. We are getting this under way as an initiative for assessing economic payoff. One thing I found in the economic studies I have seen so far is a Study A, a Study B and a Study C, but you can't relate A to B to C because there is no common thread of ground rules that allows the language to talk back and forth across those analyses.

We are planning to draw a box around and emphasize our highest payoff initiatives, including front-end research that has high potential but is not to a stage of economic assessment.

I hope I have been able to illuminate the genesis and role of the Energy Technology Centers. I was talking with Dr. Schlesinger a couple of weeks ago. As a newcomer, I thought that the ETC's had an important role and a proper role to play in the fossil energy strategy. I felt we needed to do something to refocus our energy and our pirorities, and we're doing that.

Bottom line, 'tho, our goal at PETC is to get the maximum productivity for the taxpayers' dollar in response to our mission and objectives within DOE, and in providing first-class technology support to you in the energy industry.

RECEIVED May 21, 1979.

8

Roundtable Discussions

GENERAL CHAIRMAN PELOFSKY: In presenting your economics on your various processes, I noted that you included the price of coal as an operating cost. When we talk about commercialization, we are talking about anywhere from 15,000-20,000 tons per day of coal. I would like you to comment on whether (for mines) it is proper to consider coal as an operating cost when, in fact, you will need a dedicated amount. And if you are going to have a dedicated amount, isn't it more proper to include it as a capital cost?

CHAIRMAN SEGLIN: I agree with you. It depends upon who owns the mine and who owns the plant. In all likelihood, the mine is an integral part of the plant and it would not depend upon merchant coal. But others may have individual opinions. I agree that you should include it in one package--it's a transfer cost.

PANELIST SCHMID: We feel the economics of the commercial plants in the long-range future means looking at the mine costs and using the mine as part of the capital investment, looking at the overall mine plus the plant as one entity. Generally, though, in some of the economics studies made in the past, it has been simpler just to include the coal as a transfer price and to show a plot for the selling price of the products as a function of the coal cost, serving the same purpose. If you have some feeling about what the coal cost eventually would be, this enables you to do the study with much less work and gets around the problem of having to make a detailed study of the coal mine itself, allowing the study to concentrate on the process. But I think ultimately, for commercial plants you have to look at a dedicated mine, otherwise you are probably going to be having false prices for coal.

PANELIST BLOOM: I would agree with Bruce. At this stage of analyses, we would still find it simpler to put in a price for purchasing the coal. In the case of calling out a commercial plant concept in the demonstration plant program, that was based on the procurement of coal from multiple mines. When you are talking about coals from the eastern part of the country, you are probably going to have multiple mines, and the inclusion of the economics for the mines gets to be rather complex. It facilitates

0-8412-0516-7/79/47-110-111$05.00/0

your economics to include purchased prices, and when you have multiple mines you might bear in mind that it might cost you more than if you are able to have a captive mine next to your plant.

GENERAL CHAIRMAN PELOFSKY: I agree with you wholeheartedly that it's easier to do, but it's misleading. If you are to look at the overall package, including mining, the coal gasification/liquefaction plant, that unit of operation, is in the neighborhood of 20-30%. If you don't include mining, it's in the 50-60% range. You are transferring a capital cost into an operating cost, and one of the sensitive parameters is the cost of capital.

PANELIST WOLK: It may be that the financing for both parts of that project, the mine and the process plant, will be treated differently. You can argue for government help with the process plant producing a product which will not be economically competitive until sometime in the future, but I don't think you could expect that same kind of help with the mine. I think there are depreciation credits now allowed for coal mining. Setting up the financing for these kinds of projects is much more important than arguing over whether coal is an operating or capitalized cost. It's how to get the first few plants running. I'm not sure the costs are different in any meaningful way.

HOWARD SIEGEL, Manager, Synthetic Fuels Engineering Department, Exxon Research & Engineering Company: I would like to ask Bruce Schmid a question. In the flow plan that you showed, I wondered what the initial boiling point was of the bottom stream that was recycled back to help form the coal feed slurry.

PANELIST SCHMID: That would be in the range of 850-900°F. The distillate that we take off would have an end point in that neighborhood. From our Tacoma plant experience, that appears to be about as far as we want to go and still maintain a reasonably pumpable slurry as feed to the gasifier.

H. SIEGEL: Can you say anything about hydrogen consumption in the system?

PANELIST SCHMID: Our hydrogen consumptions in our current designs will run about 4% by weight of dry coal.

H. SIEGEL: Bruce, along similar lines, can you say anything about the hydrogen treat rate that goes into the liquefaction reactor as a percent of coal feed or any other basis?

PANELIST SCHMID: That would be in the range of about 40,000-60,000 standard cubic feet per ton of coal.

H. SIEGEL: Also regarding the preparation of the feed slurry, you showed the bottom stream being recycled and mixed with the coal. Is there any other distillate stream completely free of solids that is also recycled to help form the feed slurry, or is it all bottoms recycle?

PANELIST SCHMID: We have been looking at several variations along this line. Our current thinking is that it would be basically the product slurry before the vacuum distillation but without flash liquids, and this way we can avoid adding back any distillate. We are now studying the possibility of adding back some distillate to optimize the process, but we have not firmed this up as to exactly whether we want to add it or how much.

H. SIEGEL: Bruce, did you say anything about the quantity of bottoms that is recycled relative to fresh coal feed?

PANELIST SCHMID: I can't reveal just what it is in our specific designs, but in general, it is in the 1.50-2.50 range of slurry to coal.

DONALD M. CARLTON, President, Radian Corporation: All of you talked about the need for a government role in the commercialization of this type of technology. I would be interested in what you think are good approaches for the government's participation.

CHAIRMAN SEGLIN: Well, I would think one positive approach would be for somebody high in the government to set an objective on what we are supposed to be doing. Until you set those objectives and until you sell the thing, I think you have some problems.

D. CARLTON: Some people talk about rapid depreciation of the plant, others about price supports, and others about the construction of a plant with a guaranteed customer for the amortizable life of the plant. There are a variety of suggestions. There has to be some specific government inititatives in that direction in this session of the Congress.

Some companies feel that there should be a smorgasbord type of approach while others feel there should be a specific approach. Does anyone here have a particular approach which they feel should or shouldn't be used?

PANELIST EPPERLY: Different companies prefer different approaches just because of the levels of profit, for one thing, that different companies have.

Exxon has been one of the companies to favor approaches that would reduce the impact on the capital investment (which has the greatest impact on the cost of the product), such as accelerated depreciation, investment tax credits and possibly grants that are convertible to loans after a certain period of time. All of these things have the effect of reducing the impact of the initial capital investment. But not all companies could take full advantage of all of these in a really large plant.

PANELIST BLOOM: You really have to ask the financial community what they would require in order to finance one of these plants. It's not going to be just the organization that builds the plant that will decide.

GENERAL CHAIRMAN PELOFSKY: There is one word that speaks to that, and that is "collateral". The financial people want the collateral.

CHAIRMAN SEGLIN: They want to be assured of a reasonably risk-free investment, and anything you could do in the marketplace to remove the risk would be almost imperative. I guess you do that by means of arm's-length negotiations with the government, but I don't know how you implement it.

BERNARD SHULMAN, Director of Research & Development, Tosco Corporation: In Bob Epperly's talk on the Donor Solvent process, he mentioned an operating problem: that calcium can be a problem in the form of deposits both as oolites or scale. I was curious as to what has been the experience with SRC? Do they encounter that? And how does H-Coal handle that in terms of deposits on the catalyst?

PANELIST SCHMID: In the SRC-II work that we have done at the Fort Lewis pilot plant and at the smaller laboratory pilot plants, most of it on bituminous coals, the deposits of calcium are not a problem, or at least we have not found it to be so. At Tacoma, any time we have looked at the reactor after a long period of operation, we have found no significant deposits in the dissolver or reactor. We have found some minor amounts of coke, but that's all.

The problem probably does exist with the subbituminous coals. So far we have concentrated primarily on the bituminous coals and have recognized that this will be a problem with subbituminous coals, but we have not yet tackled the solution to this.

PANELIST SWAN: We have not addressed the calcium problem. As you know, all of our operating experience has come from HRI. I am not aware of HRI having experienced any problem with calcium.

PANELIST WOLK: They have been looking for deposits of calcium processing Wyoming coal. I think there have been some very small concentrations of oolite, like structures found in the residues that have been looked at. Obviously, some of the calcium ends up on the catalyst, but it is a small proportion. The wall deposits have been checked for calcium concentration and they have been minimal. Whether that is a function of the reactor diameter using the PDU, the turbulence of the bed or really a lack of detailed observation, I don't really know. But it has not proved to be an operating problem with the H-Coal process. The period of time covered with subbituminous coal runs have been as long as thirty days.

G. H. BEYER, Professor, Department of Chemical Engineering, Virginia Polytechnic Institute: There are now design studies for SRC-I and SRC-II activities. The figures I've heard so far would indicate that probably not more than one will be supported. What do you see in the near future for the comparison between these two processes?

PANELIST SCHMID: Several years ago we made a comparative economics study between SRC-I and SRC-II. The results then indicated that the overall selling price would be about the same for the two products; namely, a liquid fuel oil of about .3% sulfur from SRC-II versus a solid solvent refined coal containing about .8% sulfur from SRC-I. Obviously, there is a substantial difference in the quality of the two products in favor of SRC-II. Also, the required selling price for a given return on investment is just about the same. We feel that this is a clear indication that SRC-II is preferable, and we've seen nothing in the past two years to change our opinion. The Tacoma pilot plant continues to be encouraging enforcing this opinion more than ever.

PANELIST WOLK: I think there have been some new developments in solid separation with SRC-I that might change the economic evaluations to some extent. However, it is informative to look back over the last year in this so-called horse race between SRC-I and SRC-II. The technical merits have not been the major issue. I think it has been the political questions. If the politics can be worked out, you might come to one decision. If they can't, perhaps there is another decision. Up until this time the political question has been very important.

LEON PETRAKIS, Senior Research Associate, Gulf Research & Development Company: Would you amplify on these technical developments that may impact on the economics of the SRC-I?

PANELIST WOLK: Since about August of 1978, we have been running a Kerr-McGee critical solvent de-ashing unit as a means of removing solids, and I think that work is very promising. The full economic implications are not yet understood. The process is still being optimized as to what it takes to get it to work and how the product recoveries are optimized. But I think it gives one another view of the overall situation in that you don't have to cope with the acres of filters question and the reliability of filters. You now have an opportunity to look at a process which is essentially a continuous one instead of a batch process.

E. L. CLARK, Consultant: SRC-I and SRC-II are being discussed, but I'm a little shocked that the only liquefaction process that is actually in operation hasn't been mentioned. No one has compared the processes that they are all working so hard on with an idealized process converting synthesis gas to liquid products. I realize that everyone can say Sasol-I is over the hill, that it is not a reasonable plant, and I agree. But everyone of these proponents has presented an idealized version that has been presented by others in a modern plant using Morgan gasification and using Morgan methods for converting the synthesis gas product to high octane gasoline.

PANELIST EPPERLY: In the evolution of technology, inevitably we get to the point of understanding very well what the problems are in the technology that we are developing. For that reason, and also because the alternative technology may also be moving, we have to review the competition from time to time. I think it is timely to go back and take another look at the Fischer-Tropsch liquids, but I don't have a quantitative answer to your question. I could tell you of the problems that you already know about with the thermal efficiency and cost. But I do think it is time that we review Fischer-Tropsch again. That's not to say I'm not sanguine about direct coal liquefaction, but I think we ought to be objective about it and consider the alternatives very carefully.

FRANK C. SCHORA, Senior Vice President, Institute of Gas Technology: Several months ago, we had this burn test at Con-Edison with the SRC-II material. Could Bruce Schmid comment on the characteristics of that meterial as compared with what you would think a general-run material would be from an SRC-II plant? There had been a comment that what was burned at Con-Edison was considerably lighter than what you would anticipate from a production run of SRC-II in a commercial plant.

PANELIST SCHMID: The material burned at Consolidated Edison was a blend of what I referred to earlier in the paper as middle distillate and heavy distillate. We find that the ratios of these two do vary considerably with different coals and different operating conditions. We made a blend with a ratio of what we felt would be representative of the material we would produce in a commercial plant. As we studied the process further, we could see that there would be some cases in which this might vary and there might be a little more of the heavy distillate to middle distillate. But the test fuel may well be representative of exactly what we would make. This is still subject to some further study and firming up of the design.

However, I think the significant part of the burning test was that the NO_x emissions were considerably less than one would predict on the basis of the nitrogen content of the fuel, and even without staged combustion, we are well below the EPA regulations. Thus, even if there is a slight increase in the quantity of heavy distillate to middle distillate in the final blend, this still wouldn't make any difference in its burning characteristics generally and would make very little difference in the NO_x emissions as well.

On the basis of this burn test, we feel that it has pretty well demonstrated that the full-range product from SRC-II will meet all the requirements of EPA on emissions.

PANELIST WOLK: We also have some questions along those lines and we are planning some additional small-scale combustion work to sort out the behavior of the medium and heavy distillates and blends thereof in the same test apparatus that was used by KVB in the original work, which used two-to-one blend.

MICHAEL WILLINGHAM, Research Analyst, President's Commission on Coal: Mr. Bloom, earlier today you mentioned the DOE criterion for evaluation systems and that not enough systems have been looked at in a standardized manner. I believe someone else also mentioned the uncertainties associated with the financial aspects of many of these processes appear to be so great that they outweigh the relative differences between some of the processes.

I was wondering whether or not a more definitive standardized analysis was in the offing and if there was anything to be gained by it, and if you could elaborate on these two aspects of that.

PANELIST BLOOM: I don't know of any effort toward expanding an effort like C. F. Braun's. We have a new contractor/monitor of gasification programs, or maybe even broader than that, in the UOP SDC contract with fossil fuel. I don't know what their emission is in this regard. The C. F. Braun studies were part of the

DOE-AGA (now GRI) cooperative effort. Therefore, they covered these processes in the DOE/AGA program and compared them with the standard Lurgi, dry bottom.

I think most of us try to look at our economics with the Braun guidelines. I think the bigger problem probably exists in the estimate of the capital costs and the capital-related costs are a large portion of the economics that go into the price of the product. I don't know the answers to that, except perhaps to have one completely unbiased, well-experienced organization do a capital cost analysis for all the processes. Even that is a problem in that process designs are in different degrees of development.

PANELIST WOLK: We have been struggling for four years now trying to do comparative cost estimates. When you get into the decisions that have to be made while you're doing a flow-sheet development, you soon discover that you really don't understand how well these plants will operate, and the decisions you're making will affect plant operability. The key thing on a capital-intensive project is what its on-stream factor is going to be. Everybody talks about a 90% on-stream factor, but no one knows how to design any of these plants to keep them on-stream 90% of the time in terms of the technical decisions you have to make or the redundancy. We need much more operating data before we put a lot of credibility in comparative economic studies. You don't go from a bench-scale study to a definitive economic comparison without a tremendous amount of risk. I don't think those should be taken very seriously just yet. We have to have some good pilot plant data on operability to know how you really have to run these processes to keep them on-line. Then there's time to do comparative economics. I feel there is just too much emphasis on it in the absence of very good experimental data, and our job should be to get the experimental data to the stage where people want to make economic decisions. Nobody is running forward to build these plants now. They are clearly non-competitive at the moment with petroleum. There's a lot of learning to be done yet and it's too early to start throwing things out on the basis of so-called unbiased and competent comparative engineering studies.

ROBERT A MOON, JR., Manager, Coal Industry Marketing & Management Department, Brown & Root, Inc.: I would like the expression of each panelist of what our marketing priorities should be. My impression is you're talking about a fuel for the utility market.

PANELIST BLOOM: I would say that COGAS talks very little about fuels for the utility market on the liquid side. The quality of the liquid product that we presently project is quoted as a No. 4 fuel oil and it is practically a No. 2. It really ought to be suitable for residential and commercial use. The naphtha we are talking about is a reformer feedstock and maybe a chemical feedstodk.

PANELIST EPPERLY: If you assume that economics will be the basis for making decisions regarding markets, I would say that there is not yet enough information available to answer the question. The work that is and will be under way will give us the basis for comparing the use of coal liquids in different markets and with the competition whatever that happens to be. I think it is prudent at this time to study a wide range of possibilities in order to get the information on which to base economic decisions.

PANELIST SWAN: Ashland is looking at refinery feedstocks plus chemical feedstocks and synthetic crude oil.

PANELIST SCHMID: We have been looking primarily at the utility markets and especially at the utilities located in metropolitan areas, such as the East Coast and the West Coast, where the restrictions on sulfur emissions are very stringent. And we see that there is a definite market here, and from our studies we feel that this is one of the earliest markets that will be viable and that we can predict will be there.

We are looking also at refinery feedstocks in other more specialized uses, over a little longer time. But we feel that the utility market is probably nearest in calendar time and opportunity.

PANELIST WOLK: I think the thing you have to remember about the utility market is that it uses the lowest hydrogen content fuels which should be some of the cheapest fuels you can make with liquefaction.

We have some real needs because we represent a segment of the country that is fairly easy to identify. If a local air pollution district wants to limit NO_x, they look at stationary sources. Some turbines in Southern California now allegedly put out about the same NO_x as a single motorcycle. I don't know if that is a true story. It may be a hundred motorcycles or something like that.

We think we are going to be a target when petroleum is taken away. I say "we" speaking for the industry, and perhaps I shouldn't do that. But I think we represent a good market for synthetic liquids, and I feel we will be an important part of any commercialization scheme.

E. CLARK: I think it is rather pointless at this time to argue about which is the biggest market for anything. But if one looks fundamentally at what one can do, one would have to say that for the utility market probably medium-Btu gas is the ideal fuel because it may be less expensive than liquids. It obviously doesn't have the storability that is important to the utility. But I have always looked at the liquid market as a market that can't be supplied by any other source, and I always think of it as a transportation fuel market. The only purpose for making utility fuel is that hopefully some day you can convert to gasoline. And to say that utility fuel will benefit from the low sulfur content of synthetic liquid fuels isn't exactly true. It doesn't compare with the low sulfur content that you can achieve with gasification, again warranted that you would like to have a storable fuel for feedload potential.

This is one of the reasons I brought up the liquids from synthesis gas as an ideal combination for the utility which would provide the methanol that is needed for the tip of the peak, and also possibly an ideal hydrocarbon turbine fuel which you will never get without a great deal of processing from coal through hydrogenation.

PANELSIT WOLK: Zeke, I think I should have said, if I didn't, that we are looking not at low sulfur fuels but low hydrogen content fuels.

Let me also state that I said earlier this morning that as for the baseload market we think that medium-Btu gas plus combined cycles is very competitive. For the intermediate and peakload market, though, we think that storable liquids are important. We are also looking at some variations on the theme that you proposed of diverting part of that intermediate Btu gas into methanol and using that to meet the peaks. Those schemes are under investigation.

GENERAL CHAIRMAN PELOFSKY: I have another question, Len, which I will address to everyone on the panel.

What about the environment? How does RICRA affect you? What about air and water quality?

PANELIST EPPERLY: We think that we can meet the standards of 1985 as we understand them, assuming something unpredictable is not going to happen in the next several years. The problem comes down to money to meet them, and also the time required for permitting. I don't mean to minimize either one of those things. It will cost a lot of money to meet the 1985 standards, but it can be done.

GENERAL CHAIRMAN PELOFSKY: I have a feeling that in your commercialization schedule you have things such as detailed engineering, construction and operation, but too infrequently do I see time allocated to such things as permitting, EIA's, EIS's and so on that maybe should precede the detailed engineering phase of a project.

CHAIRMAN SEGLIN: That is why I mentioned, Arnold, that ten years is the lead time on these projects. That might be too short but a realizable lead time for products of this nature, without all those regulatory hazards, is considerably less than ten years.

PANELIST EPPERLY: I would like to hear Bruce comment on that because I think they have the fastest schedule right now.

PANELIST SCHMID: Yes, I did want to make a comment on this. The environmental situation doesn't get mentioned too often. That is true because we tend to concentrate on the process a little bit more. But we have not overlooked it at all. In fact, we have already started baseline studies near the Morgantown, West Virginia, site for the demonstration plant. We have been gathering data there for several months, and we have a meteorological tower erected in that vicinity. We have devoted a considerable amount of time and effort to gathering the kind of baseline data that we will need to assess the environmental impact of this plant on the surrounding area. The study of the environmental effects is one of the key parts of the demonstration plant program. When we are finished with the demonstration program, we should have a very good knowledge of the environmental effects as well as a knowledge of all the scaleup factors that are involved in the development of the process.

PANELIST BLOOM: The same is true in the high-Btu gas demonstration plant program. An integral part of the work under the contract is the environmental assessment. There are about two tasks assigned to that. It started from the beginning of the contract period because we had to come in with a site. And it will lead, of course, to providing the information for the Environmental Impact Statement.

In addition, in the demonstration plant program, the need for obtaining the necessary authorizations and permits was recognized, and this was made part of the early part of the program. So when you say to do it before the detailed engineering, actually this is the way the program was laid out.

Of course, on the other side of the fence, you have to be continually concerned with the environmental problems that the process creates and keep these meeting the standards. Another factor is the difficulty of scheduling into any plan the lawsuits that you may well run into when you start to build one of these plants anywhere.

In our case, I mentioned possibly starting a commercial plant project in 1986, if everything goes along according to the latest schedule. This was on the assumption, for example, that it would be a first commercial plant at the site of the demonstration plant with most of the environmental effort already taken care of. If you don't have that situation, then in our thinking we do schedule in a certain amount of time. I can almost guarantee we will never schedule as much time as it probably will take.

RICHARD A. PASSMAN, Director, Coal Resource Management, U.S. Department of Energy: Most of your processes produce a significant amount of high-Btu gas, and I wondered how you treat it economically and regulatorywise. Will you be required to file with FERC or are you going to sell it to a transmission company at a given rate? And in your economics, what did you assume to be the value of that gas?

PANELIST SCHMID: In all of our economics, we have calculated the required selling price for the total Btu output of the plant and presented the economics this way. This is admittedly an oversimplification, because it does not take into consideration any possible price differential between the gas and the liquids. However, we feel that a differential is justifiable. The gas is undoubtedly more of a premium product than the liquid, and without a controlled market there certainly would be a price differential. Exactly what this would be is open to question. This is really yet to be tackled. There are some ways of trying to estimate this. We have made an attempt to estimate this. But it is uncertain enough that we have not attempted to put this into our formal economics yet.

PANELIST EPPERLY: In our base case design for EDS, we don't make gas as a product. We maximize liquid yield. The options that I talked about involving using combinations of processes for bottoms would look attractive if gas could be sold in parity with the liquid fuel. But there is clearly some uncertainty regarding exactly how all that would work out. So that is the reason that in our base case we make no gas.

FRAN R. CONNOR, Research Assistant, University of Colorado: For Bruce, please. In your final product, the SRC-II, what is the ratio between the fuel oil to pipeline gas?

PANELIST SCHMID: It depends a bit on whether you include the LPG products in the gas or not. If you include LPG in the gas and include the naphtha in the liquid, it is probably something like a two-to-one ratio or three-to-two ratio, with the liquid being the greater.

F. CONNOR: What is the relationship, if there is one, between the environmental studies at Battelle Pacific on SRC and your two Fort Lewis and Tacoma plants? Is there any tie-in?

PANELIST SCHMID: They are not directly related. We have had extensive studies done in our Tacoma pilot plant at Fort Lewis, Washington. These environmental studies involve both in-plant studies and studies of the atmosphere surrounding the plant. These are quite extensive. We are doing this as part of our pilot plant program. I am not familiar with the Battelle studies, but I know of no direct connection between them.

H. SIEGEL: I have a question for Jack Swan. Jack, in your description of a conceptual 20,000 ton-per-day commercial H-Coal plant, you mentioned that the plant would include a number of parallel liquefaction trains. Can you say approximately how many?

PANELIST SWAN: We have just recently embarked upon the commercial design study, and we are looking at possibly ten trains. But, admittedly, that is very rough at this time.

H. SIEGEL: Thank you, Jack. I have one more question for you. Previously you mentioned that the main goal of the H-Coal process was to produce synthetic crude. Does that mean that you think that the 400 - 1000°F material in that synthetic crude would actually be converted in a refinery to other products?

PANELIST SWAN: Yes.

H. SIEGEL: In other words, it might go to hydrocracking or cat-cracking?

PANELIST SWAN: That's true.

D. CARLTON: The environmental discussion suggests that any of these plants are going to have to be sited away from the end user because of PSD considerations. Do any of the economics presented this morning include product transportation?

PANELIST BLOOM: The ones I presented were generalized plant tailgate prices. You know you can put in any figure for transportation because you just don't know how far you are going to have to pipe it or transport it to your customer. That might be one point where I think we are fairly uniform, except in special cases.

GREGORY BOTSARIS, Professor, Department of Chemical Engineering, Tufts University: I have a question for Mr. Wolk. The question concerns coal/oil mixes which I think, at least indirectly, are part of the coal dilemma. The question is: What is their thinking about the potential of coal/oil mixes?

PANELIST WOLK: I am not directly involved with the work on coal/oil mixtures. I know we are sponsoring some work and intend to run some utility scale tests. But my own view, which may or may not be consistent with EPRI's view, is that if you try to put a coal/oil mixture into a boiler that is an oil boiler, you have a whole set of ash problems to cope with which makes life very difficult. It may be useful to do it in a boiler that was originally designed for coal and has been converted to oil and now you want to go partway back. The dilemma on converting back, though, is that many of the off-sites that you need to handle coal are not available at a lot of utility stations where coal was formerly burned which were then converted to oil. You need rail sidings to bring in that coal; you need all kinds of conveying and handling and crushing equipment to deal with it. I think when utilities, especially in cities, were converted over from coal to oil, they got rid of all that stuff. Now if you want to make the move back, you just can't do it. The application of coal/oil slurry, I think, is limited to places where there are those facilities and to boilers which are capable of handling a bottom ash.

G. BOTSARIS: What about the central activities which can always use the present transportation units to carry the liquid fuel now?

PANELIST WOLK: But this liquid fuel now is a slurry of coal and oil. You have the same problem with the generating station dealing with the coal ash that is now contained.

R. MOON: I have heard the word "commercialization" this morning and this afternoon. I have heard the word "economics." I would like to pose a question to the panel members who are the representatives from industry. What kind of incentives do your parent companies need to commercialize the technologies you have been talking about? Specifically, what do you need from the Federal Government?

PANELIST BLOOM: It's really the financial people who have to answer that question, not the technical people. Wouldn't it be nice if it would be money that didn't cost you anything?

R. MOON: That's part of the dilemma.

PANELIST EPPERLY: I think we know the incentives required are large, and this immediately raises certain types of political questions as to whether some types of incentives are more possible than others. It will be very, very difficult to answer the question except based on detailed discussions with the people who will be able to provide the incentives.

L. PETRAKIS: Can you apprise us of what is the current international interest in the various competing processes especially among the Japanese and the Germans? And is that likely to be a factor as to which one of these competing processes might get most of the federal dollars?

PANELIST EPPERLY: Well, I can comment on EDS. We have $20 million from a group of twelve private Japanese companies. The Japanese government is also going to put some money into the program, but they will be in a minority position. I think it is quite significant that the private industry in Japan came up with $20 million. In addition, Rural Coal has joined our project at the $5 million level.

PANELIST SCHMID: I might add that there is considerable interest today in the SRC-II process in both Germany and Japan, and this should certainly help in the development of the SRC-II demonstration program. But that is about as far as I can go in commenting.

A. CONN: On the question of the financial aspects, I think you are making a mistake in referring this to the financial community, because every comment I have heard from the financial community is that they would like to have something that is tried, proven and ready to operate, and they guarantee a certain amount of return. It seems to me that the financial community is going to duck out on this very quickly, and the whole thing goes back to the government having to do something to back it up.

I would like to ask for any further comments on that point because I don't think the financial community is going to take any of the risks that we are talking about.

CHAIRMAN SEGLIN: Zero risk corresponds to zero profit.

PANELIST SWAN: That isn't exactly what I meant by going to the financial community. The financial community really has to state what it is they would accept in order to be in the position of providing the funds for the construction of plants. I agree with you. They are not the ones, unfortunately, to whom you can look at this stage without a lot of government support or industry putting up all its assets, which I doubt it is going to do. But we can talk about loan guarantees, grants, price guarantees, or whatever the people you are going to borrow your money from feel is acceptable.

A. CONN: Bob McNeese made a comment on how to handle caking coals. He left it up in the air. I was hoping he would be here this afternoon to talk about it. I don't know whether you are in the position to comment on what he might have said or not.

CHAIRMAN SEGLIN: I think one of the options he gave was the Westinghouse gasifier, the use of a type of backmixed dilute phase system for mixing the fresh coal into the hot bed. That's known technology. At least, it is published.

The thing I cannot comment on is what he was referring to with regard to the proprietary developments from Carbide. They apparently did some work in the course of the Coalcon process. But I think, as I remember, they still wanted to pilot the process. The demonstration plant project did not include piloting.

A. CONN: So we don't know whether they have really solved the problem yet or not.

CHAIRMAN SEGLIN: The rumor is that they were successful on a small bench scale or pilot scale. But I don't know how successful.

A. CONN: The other question I had for him had to do with the pumping or getting solids into a high-pressure reactor. I was hoping to hear something about the developments in that area.

CHAIRMAN SEGLIN: I think their conceptual design was based on lock hoppers. He alluded to slurry pumping, but that's a horror.

A. CONN: If anybody has had any experience with lock hoppers, I think they would rather have almost anything else. There had been talk of something like extruders that work on plastics and actually force the material into some type of a solid that can be injected. But I guess we don't know about that either.

CHAIRMAN SEGLIN: We can make the same comment about the extruder. I think that could be a horror, too.

GENERAL CHAIRMAN PELOFSKY: Jack Silverman is in the audience from Rockwell. Maybe he can answer it and speak to the problem.

JACK SILVERMAN, Director, Fossil Energy Conversation Systems, Rockwell International Energy System: I guess for the process that we are pursuing, we view it as a two-step type problem. One is to get pulverized coal up to pressure in a feeder, and the second step is simply to use pressurization with appropriate attention to detail in tanks, lines and valves, but use pressurization to move the solid just exactly as you are moving a liquid. We have been successful in Step 2, which is the only step that we have addressed, in moving a solid that way at various rates, at least up to several tons an hour.

We have been following with great interest the DOE-sponsored developments in the various so-called pump programs, and we hope that perhaps one of them will come up with a system that will take pulverized coal from ambient pressure to high pressure of several thousand psi. Lockheed, for example, I understand has a system that will go to at least 600 or 700 psi depending on the pressure.

PANELIST WOLK: I wonder if I might presume and ask Bruce a question. I know Gulf has done some interesting work at Tacoma on hot slurrying of rather coarse particles. I wonder if you could share that with us.

PANELIST SCHMID: You may be referring to the work on extrusion. We have been looking into the possibility of extruding the coal with a small quantity of liquid into the slurry mixing tank. Basically, this accomplishes the initial wetting and mixing of the coal, and then the mixing is carried on further in the mixing tank itself. This is a system merely to get the coal wet with the slurry, which is no easy problem.

The effort here does look encouraging, and this system may well be another alternative that would look good for the demonstration plant design. We are testing this out now at Tacoma. At the present time, it is not a part of the demonstration plant design. At the same time, we are continuing our tests using the high-speed mixer at Tacoma. This work is also encouraging. It gets the coal wet and mixed initially, and we can keep the coal suspended and mixed by continually circulating it around the loop and back into a larger mixing tank.

So, basically, we have two alternatives now for slurry mixing and pumping. Of course, the entire mixture is then pumped to the high pressure necessary for the reaction by reciprocating pumps. And we have good experience at Tacoma with the reciprocating pumps. This is one of the aspects of the process that has worked fairly well and one of the more successful operations in the plant. We view this as being one of the problems that is less of an uncertainty than some of the others that I mentioned this morning.

We are also looking down the road at the possibility of centrifugal type pumps for this high-pressure application. But we haven't gone far enough along with this to feel that we could design this into the demonstration plant. So, at the present time, it is reciprocating pumps and probably will be for some time yet.

H. SIEGEL: I would like to come back to this question of incentives for a moment. I think there are two basic approaches that industry could take to this issue. I think one of them has a much better chance for success than the other, and I would like to describe what I mean.

The first basic approach would be to promote the idea for the government to put in place a series of possible incentives (I believe someone this morning called it a smorgasbord) from which individual companies could choose for individual projects. Personally, I believe that the chances of this happening are pretty slim, because politically it would be difficult. There would always be the concern on the part of the government that this would provide a route whereby individual companies with individual projects could be provided with more incentive than they really need and, hence, could obtain a windfall on their particular project.

The other approach which I believe has a greater chance for success is for individual companies or groups of companies to take upon themselves the initiative and the responsibility to formulate individual, commercial synthetic fuel projects, to calculate their economics, to define the particular incentives they would need in order to go forward with their particular project, and then to go to the appropriate government agency and request those particular incentives for only that particular plant.

That, to me, has a real chance for success because that is not a set of general incentives that a lot of people can possibly take unusual advantage of. Instead, it is a project-by-project basis, and if done properly, I am not sure what basis the government would have for not granting the particular incentives for that particular plant.

CHAIRMAN SEGLIN: I would like to hear some comments from people who represent the government. I don't think the panel can address themselves reasonably to that.

RICHARD A. PASSMAN, Director, Coal Resource Management, U.S. DOE: I am going to address some of this tomorrow in the session. But, in actuality, the government is not in a position of building a capacity of a particular size in the country. However, we are interested in demonstrating the capability, generating an experience base. I think the latter approach is a good one, but rather than saying it must be provided to everyone, it might be that a particular circumstance of a particular organization in the plant could generate that single experience base at a lesser cost to the government because it's an add-on to something that they have had or it's a particular situation that will ensue, but still give an experience base of capital operating cost. In one case, it would be the handling of coal, the handling of ash, the environmental conditions, and so forth.

So if you don't consider it a broad application for everybody, but instead a single demonstration of a type, be it direct liquefaction or indirect liquefaction process, I think this is a very likely thing and something that we intend to pursue.

ROBERT P. SIEG, Manager, Synthetic Fuels, Chevron Research Company: I was just thinking of an example of this type. It's not on the coal end but in the x-32 shale retorting end, and Union Oil proposed a $3 valve tax credit at which point they said they would be willing to use their own money to go commercial. What happened, of course, is that they came back and said, "Well, maybe on the first five or ten thousand barrels and then maybe for one year." That isn't what they had in mind.

CHAIRMAN SEGLIN: I think the message has to filter somehow down to the government that private industry is profit-oriented, and I don't think anybody in private industry is about to commit hari-kari with their company. We're talking about the big bucks and with no expected return for the dollars that you are referring to. Maybe I missed the point.

R. PASSMAN: I think the suggestion was made that each company decide what it needs in order to go commercial. That could be a capital grant or an investment tax credit or whatever it might be. But if, in their own calculations, that gave them a requisite return on investment or a discounted cash flow of some form that they needed and they received, because of all those that came in, that happened to be the best deal for the government then there was a match and therefore they ought to proceed with it. If they lost their shirt or other parts of their anatomy, it would be their own fault.

W. C. LANNING, Project Leader, DOE, Bartlesville, Oklahoma: One question for Mr. Epperly. I was interested to hear some comments a little while ago about small refined or refining feedstock liquids. That is what we are more interested in at Bartlesville.

About a year ago, you mentioned some hydrotreating of this full range of heavy distillate in doing some catalytic upgrading, and you mentioned that there were some plugging problems that developed in small bench-scale work, I believe, in from one to five days. At that time, you were looking for the possible source of that trouble. I haven't seen reports since then. I wondered if you have turned up an answer as to what caused that plugging or a solution that you could talk about.

PANELIST EPPERLY: We think we found the answer to that problem. We believe it was an experimental problem in the unit, having nothing to do with the material that we were treating; but I am not yet in a position to absolutely confirm that since additional testing is under way.

W. C. LANNING: Good. Another question or comment, perhaps getting into more research aspects. It's a bit speculative.

We, of course, are interested in lighter liquids, perhaps transportation fuels, as Zeke Clark mentioned. I made a pitch at a meeting, I believe three years ago, about the possible use of subbituminous coals as a compromise feedstock for making lighter liquids. In the three or four years I have been looking at this problem, just recently considerable bits in the literature indicate that they might be reasonable feedstocks because they make less complex liquids and they are easier to upgrade. We are not in the coal liquefaction business, but we did make some batch preparations primarily for characterization purposes, and we found that in the first place, the lower ranked coals are much more reactive. This does require some catalysts. These were batched primarily for hydrogenation. The subbituminous coal, for example, reacted rather vigorously by laboratory standards, as much as 200°F below that at which the present processes are operating. The crude liquid produced then was much more easily upgraded to a given low of nitrogen of something like .2% as a possible feedstock than those of higher rank, indicating that the complexity of the compound was much less. The nitrogen was easier to remove.

The indication seems to be that to exploit these lower ranked coals, one needs more reaction temperatures for the initial reaction because the Bureau of Mines, ten years before General Brill's reference to the start of the work back in the late thirties, found that the lower ranked coals were very reactive, actually too reactive. Mr. Epperly, I believe, mentioned that the subbituminous coals are more difficult to process in their process. This apparently was because there are thermal re-accomodations which take place and you make worse products than you started with in some respects, a lower temperature catalytic reaction. And work at the University of Auburn indicates that these coals fall apart almost instantaneously in the Donor Solvent at as low as 350°C. This might be 650°F. Once it is liquefied, then the material could be subject to upgrading by more conventional catalytic upgrading.

Would any of you have any comments on the possibility for research along these lines?

PANELIST EPPERLY: I'll start. First of all, with Illinois coal we can vary the amount of light material, meaning below 350°F boiling point, from roughly 25-55% of the total liquid product. We think their flexibility one of the advantages of the Donor Solvent approach.

Turning to the question of Wyoming coal or subbituminous coal, I mentioned in my talk that we have actually identified three ways of increasing the liquid yield in the process. For example, one of them involves using a low temperature in the first stage of liquefaction to take advantage of the fact that some of the materials are more reactive and to liquefy those and stabilize them prior to completing the liquefaction at a higher temperature. That does work. It does increase liquid yield.

I would say that we have not had experience similar to yours with regard to making materials from subbituminous coals that are easier to upgrade. That may be because we control the conditions very carefully to avoid forming large amounts of gas which, as I indicated, in the base case we don't want to make as a product. Rather, we maximize liquids and only have the amount of gas we can use internally in the process.

Under those carefully controlled conditions, we do have the high nitrogen levels that I showed in the products, and as far as the upgrading is concerned, it's the nitrogen that controls. The sulfur is relatively easy to remove and the nitrogen is relatively difficult to remove. One of my slides showed that the nitrogen content of products from the Wyoming coal is about the same as from Illinois coal or from the lignite from Texas.

So I really can't confirm what you say, and I can't explain why the observations are different.

PANELIST WOLK: It is certainly true that subbituminous coal reacts at very low temperatures, but what happens at those low temperatures is that the solvent you use is incorporated in the coal structure to a fair degree. You don't get it back. You have a process that is going to run downhill very, very quickly. We have done a lot of work with unhydrogenated solvents on this and found that incorporation is a major problem. We, too, have been looking at lower temperature reaction conditions as part of our work. We think that is an objective that would have a profound impact on economics. We just haven't gotten there yet for sure.

W. C. LANNING: You would, of course, have to have a second stage for further upgrading, I am sure. There is essentially a second stage required in all these processes if we are talking about going to refining feedstocks. Certainly I would agree, from my limited knowledge, that the low temperature at one stage would not do it.

CHAIRMAN SEGLIN: Could you have a compensating factor if you used catalysts rather than non-catalytic systems?

W. C. LANNING: Surely some kind of catalyst would be needed at a lower temperature. The reaction alone probably would not do it.

CHAIRMAN SEGLIN: The SRC is non-catalytic, isn't it?

PANELIST EPPERLY: I think, if we understand this, the primary variable here is the amount of hydrocracking that takes place. The more hydrocracking you get in the process, regardless of the conditions, and I know this is a little bit of an over-simplification, the more nitrogen bonds, sulfur bonds and oxygen bonds you will attack. Hydrocracking will not only lower the molecular weight of the product, in some cases into the gas range instead of gasoline, but it will also just make the products cleaner in terms of sulfur, nitrogen and oxygen levels.

PANELIST SCHMID: The only comment I might make with respect to the catalyst is that, of course, there is a catalytic effect in SRC-II from the mineral residue. It is not only the mineral residue in the coal feed itself, but also in the recycle slurry. This increases the concentration of coal minerals in the reactor considerably and adds to the reactivity of the system, and we get a greater conversion because of doing this than we would otherwise.

CHAIRMAN SEGLIN: There is a possibility you can tailor a catalyst better than having it inherited.

PANELIST SCHMID: Sure. There is always a possibility of making a better catalyst, but we have found in our studies that the catalytic effect we do get from recycling the coal minerals is sufficient to do the hydrocracking job that we need to do, and I think that is the important point.

R. SIEG: Nobody has yet mentioned the incremental cost of the added hydrogen required to work on lower-ranked coal, the subbituminous coal, to go all the way to transportation fuels. This will require a substantially larger amount of hydrogen, and this will add to the cost of the liquid product and may very well be more important than the higher reactivity of the subbituminous coal.

CHAIRMAN SEGLIN: Of course, part of that equation is the relative cost of the coal. Hopefully, the western coals will be less costly. It is a trade-off. I don't know, though, what the answer would be.

PANELIST EPPERLY: I don't think it is yet clear to us what coal will be optimum because there are conflicting factors at work. Of course, coal that can be surface-mined is much cheaper as it goes into the plant. But, on the other hand, the surface-mined coals do have higher oxygen contents, they require more hydrogen and, as I mentioned, they have some other problems, such as the calcium carbonate formation. I believe considerably more work will be required before we will know whether there is an optimum coal.

CHAIRMAN SEGLIN: And the transportation costs usually would be against them in the marketplace.

PANELIST EPPERLY: I think the main point is, if you really believe that there will be a large coal liquefaction industry some day, one of the main advantages of coal is that it provides the opportunity to spread the environmental costs, whatever they are, to different parts of the country. This advantage is something not available to us with shale oil. In our work, we are really trying to focus on a technology which can be applied to a broad range of coals just for that reason.

CHAIRMAN SEGLIN: You have to use them all, don't you?

PANELIST EPPERLY: Eventually.

A. CONN: I would like to go back to this question of the catalytic effect of the solids in SRC-II. I understand that it does take some time to build up the amount of solids needed to get the conversion, and I was wondering whether you could tell us anything more about how you plan to do this in a large plant. It sounds to me a little bit like the problems in an H-Coal reactor where good contact must be assured between the catalytic solids and the incoming liquid and gas. I was wondering if you would care to comment on the design to accomplish that in a large reactor.

PANELIST SCHMID: It does take some time, but the time is probably on the order of just a few days. It is going to take this long to get the thing lined out during the start-up anyway. So by the time you have it lined out, the reactor is going to have sufficient coal solids in it to do the job.

A. CONN: Then how do you assure a good contact between the incoming liquids and these solids that have to build up someplace in the reactor?

PANELIST SCHMID: Well, we do this partly by where we add the hydrogen and how we add the feed, and by the extent of back-mixing that we get in the dissolver. All of this adds up to a system which really gives us very good contacting.

CHAIRMAN SEGLIN: Are we in the position of the auctioneer who is about ready to sell the prized object saying, "Going, going"?

RECEIVED May 21, 1979.

COAL GASIFICATION

Section Introduction

CHAIRMAN CONN: Whereas the first part of this book dealt with specific processes, the second part deals more with the philosophy and overall management. The reasons for this are two-fold: one--the ability to gasify has been demonstrated many times over, both in this country and in other places, and so the question is why isn't it commercial at this point; and two--we have so many different processes in coal gasification that it would take two days just to cover them. So with these two things in mind, we are going to talk more generally today.

As I see it, one of our national probelms is that our politicians and industrial leaders often tend to concentrate too much on the near-term without enough concern for the long-range future. Your presence here today indicates that you do look at the long term, and we recognize that our proven reserves of coal far exceed our gas reserves and that sooner or later we will have to make gas from coal. Up until recently, gas has been scarce, and yet we have seen news articles saying that there is plenty of gas, in fact, even an over-supply. We have seen articles in Fortune and in the National Geographic that have featured the geopressure zones under Coastal Texas and Louisiana. Early estimates were three thousand years' supply of gas at current consumption rates. More recently, the Department of Energy urged reduction in the use of liquid petroleum by substituting natural gas. So why in the world should we be worrying about gasifying coal at this time? This is one of the sources of our dilemma. I don't have to tell you that some of these recent articles are very "blue sky". For example, research into the geopressured zones has shown that there are difficulties not only in deep drilling, but also in the handling of tremendous amounts of water, the large amount of dissolved solids in the water, possibly even more than in sea water, and so corrosion will undoubtedly be a problem.

In recognizing these trends and problems, it seems that we still should continue to develop our plans for making gas from coal to assure continued availability of gas for many current uses. When the big snowstorm hit Chicago this winter, I realized

0-8412-0516-7/79/47-110-133$05.00/0

that if the gas were shut off we would have absolutely no way of heating our house. So for such necessities as home heating, we certainly need assurance of continued availability of gas. And we also want to make greater use of coal to decrease our imports of crude oil and help guarantee national security by making the United States less vulnerable. Finally, if at all feasible, we need to establish a ceiling for the cost of energy to improve our bargaining position. That last one may just be a hope, because it seems that as the price of the natural products goes up, the synthetics seem to go up right with them.

9

The Future of Coal as a Source of Synthetic Fuels

GERARD C. GAMBS

Ford, Bacon & Davis, Inc., 2 Broadway, New York, NY 10004

The title of my paper "The Future of Coal as a Source of Synthetic Fuels", might better have been stated as "Is There a Future for Coal as a Source of Synthetic Fuel"? In which case, my answer would be that if Washington continues on its present incredible suicidal direction of destroying our domestic energy base, there is no future for coal for synthetic fuels and in fact, there is no future for synthetic fuels from coal or from anything else. This is particularly true for the next 10 to 20 years - after that, there could be a future for both.

It should be kept in mind that synthetic fuels from coal are but one part of an overall very complex mixture and interrelationship of energy supplies and energy demand.

For example, the current policy in Washington to keep the energy demand tuned to a zero growth economy - 2% per year or less - means that conventional sources of fuels, particularly coal and natural gas, are currently available in excess and will in the future be available for a number of years longer than previously anticipated. The restraints on the economy will in turn mean that synthetic fuels from coal will not be produced commercially for many years to come.

If energy requirements are actually higher in the future and growth is on the order of 4% to 5% per year, the unfortunate result is that the increased demand will be supplied by imported oil. This could be met by synthetic fuels from coal if the plants were in place, but none is in place and none is going to be in place for many years to come.

America is slowly but surely being destroyed by a collection of groups and individuals who claim to be pursuing their objectives out of concern for the well-being of this country. Unfortunately, if they are successful, it will result in the collapse of the United States as an industrial nation. Dr. Peter Metzger, Administrator of Environmental Affairs for Public Service Company of Colorado referred to these groups as "Coercive Utopians" in a

0-8412-0516-7/79/47-110-135$08.75/0

recent speech in which he described the capture of the wealth-generating machine of society--what we call the economy today--by people in and out of the government who want to turn it off!!

As a result of actions taken by many groups and individuals we are producing less and less domestic oil and natural gas, we are producing less coal and our ability to construct and operate more nuclear plants is being prevented by many groups and government agencies. At the same time, we are importing over twice as much oil and paying ten times as much as we did six years ago. Congressional actions in the past few years have made it certain that the United States will not only be consuming more oil than it should be, but that it will be uneconomic to produce more domestic oil, and, as a result, we will be importing more and more foreign oil to meet our requirements for many years to come. Our costs for energy have grown from 2 percent of our GNP to over 12 percent of our GNP in less than 6 years. Our failure to recognize the role that energy costs are playing in world economics is preventing us from solving our domestic problems.

Since 1972, the source of the energy used for producing our GNP which has been based on imported oil has doubled, increasing from a level of about 13% of our GNP in 1972 to about 26% of our GNP in 1978.

The Carter Administration appears to be in the process of changing its energy policy on a daily basis with hardly any way to predict which way it will go on any one day.

In November 1978, Secretary Schlesinger of the Department of Energy stated that all industry would be forced to switch to coal from natural gas and oil. However, shortly after, the DOE Secretary Schlesinger announced in December 1978 that the government now wanted electric utilities and industrial plants to switch back to burning natural gas rather than burning oil.

This latest twist in the Energy Department's sometime confusing fuel-use policy emerged according to the *Wall Street Journal* (January 15, 1979) when the DOE ordered Public Service Co. of Colorado to stop using gas and start burning coal at three of its power plants.

"This order was issued just days after Secretary Schlesinger publicly restated the administration's *new* gas policy: If they can, utilities should burn gas to avoid the use of oil, to use up an expected short term surplus in the domestic natural gas market and to provide continuing incentives for domestic gas production. Mr. Schlesinger said the agency would exempt utilities that can burn gas from provisions of the 1978 Coal Conversion Act requiring a switch to coal."

"Prior to the emergence of the new policy in the two month period, December 1978-January 1979, the Carter energy program and the 1978 Coal Conversion Act (passed in November 1978) has treated gas, along with oil, as a scarce fuel, and called for replacing it with coal or other alternate fuels as much as possible."

"But immediately after the passage of the five-part Carter National Energy Plan in November 1978, it became apparent that a national gas surplus was developing rapidly due to Congress reshaping the original Carter proposed legislation to allow much higher prices for gas producers. The original Carter policy was based on the erroneous premise that higher prices for natural gas would not produce greater supplies of natural gas. It only took a few weeks after the passage of the National Energy Plan I for natural gas to become a surplus "scarce" fuel."

"Hence, the sudden reversal of the Carter Administration on their energy policy and their new program of encouraging the use of natural gas for the next three to five years and the promise that exceptions would be made in the coal conversion policy."

"To say that electric utility and industrial plant managers are confused by the latest flip-flop in the Carter energy policy is to put it mildly."

"In issuing the previously cited order to Public Service Co. of Colorado to go off natural gas and switch to coal, the DOE added a new twist to the seemingly contradictory policy. It said that its recent statements about encouraging gas use applied only when the chance of power-plant fuel was solely between oil and gas. When coal or other fuels are readily available, as in Colorado, the DOE said it still intends to push for coal conversion."

"DOE officials keep insisting that their latest change in policy is part of an effort to curb the level of oil imports. By ordering the Colorado utility to stop burning gas and switch to coal, they explained, more gas would be freed for use as a replacement for oil by other utilities that are unable to convert to coal quickly."

"Thus, DOE said, utilities will get different treatment, depending upon the availability of coal. If a utility burning gas or oil can readily convert to coal, it will still be ordered to do so. But if conversion to coal isn't practical in the next few years for utilities currently burning oil or gas, they'll be urged to keep using gas or to switch to gas from oil."

"In the longer term, the DOE said, it still wants utilities to use coal and other alternate fuels, instead of oil or gas--and new plants will be discouraged from using gas."

"The order to Public Service Co. of Colorado to switch from gas to coal may have more symbolic policy impact than practical importance. Energy Department officials concede that utility was gradually converting to coal from gas anyway, and the coal-conversion order won't take effect unless federal and state environmental agencies approve it after review process that could take a year or so." (1)

The new DOE policy of pushing the use of natural gas received prominent notice when Secretary Schlesinger of DOE gave a speech on January 9, 1979 in New York City to the National Association of Petroleum Investment Analysts and The Oil

Analysts Group of New York. Secretary Schlesinger described his speech as a "non statement of Policy."

The principal message which he gave in his speech was that electric utilities and industrial plants should in the short run (the next three to five years) turn to natural gas in their existing plants instead of to oil.

Dr. Schlesinger maintained that the Carter Administration remains committed over the long run to the use of coal instead of oil or gas in new boiler facilities.

Source of gas, other than natural gas production from the lower 48 states, were arranged by Secretary Schlesinger in the following hierarchy of decreasing marginal attractiveness:

1. Alaskan natural gas
2. Canadian pipeline gas
3. Mexican natural gas
4. Short-haul LNG
5. Domestically produced synthetic gas depending upon resolution of technical problems and cost
6. Finally, at the end of the line is long-haul, high-cost, possibly insecure LNG

Secretary Schlesinger ended his speech with the following message:

"In the near term we have a gas surplus. Until we find ways of effectively utilizing that surplus, we are under no pressure. In the longer term gas prospects are relatively attractive, far better than the prospect for oil. Overall, supplies are prospectively adequate. Indeed over the next 20 or 30 years, gas usage may well rise here in the United States. Above all, we must recognize--as we failed to recognize before the passage of the Natural Gas Policy Act--that we are under no immediate pressure. We have the opportunity to develop our policies intelligently, as uncertainties about domestic supply are reduced, and as our understanding about prices and availability of alternative supplies is enhanced."

The natural gas supply picture has shown improvement during the past few years as shown in the following table from the New York Times of January 14, 1979.

NATURAL GAS DATA - TRILLION CUBIC FEET

PRODUCTION	1974	1975	1976	1977	1978
Domestic U.S......	20.7	19.2	19.1	19.2	19.0
IMPORTS					
All Sources.......	0.96	0.95	0.96	1.1	1.0
CONSUMPTION					
All Uses..........	21.2	19.5	19.9	19.5	19.8
PROVEN RESERVES					
U.S. & Alaska....	237.1	228.2	216.0	208.8	203.1

The Carter National Energy Plan I contemplated that coal production could be doubled between 1977 and 1985, while at the same time, the Clean Air Act was made even stricter, the Federal leasing laws were tightened and a Federal Strip Mine Law was enacted and the Hazardous Materials Act was published for comments. Again Washington has lost all sense of reality since it will be impossible to produce coal at the level which the Carter Energy Plan has projected--and the ironic part is that the restrictions on its mining and use have been forced on the coal industry by Washington itself.

A brief tabulation of some of the estimated additional costs which the coal industry faces as a result of recent legislation and regulations shows that coal is rapidly being priced out of the marketplace. Examples of these new additional costs are as follows:

		Additional Cost Per Ton Coal		
1.	Surface Mine Regulations	\$6.52	to	\$17.17
2.	Hazardous Materials Act	\$5.00	to	\$10.00
3.	Stack Gas Scrubbing (Capital)..	\$15.00	to	\$18.00
	to remove 90% of SO_2 (Operating) with high sulfur coals	\$15.00	to	\$30.00
	SO_2 Removal Sub Total	\$30.00	to	\$48.00
	Total Additional Costs Three Items Listed	\$41.52	to	\$75.17

It will be impossible for coal production in the United States to reach the Carter Plan's projected level of 1.265 billion in 1978 and an increase of this magnitude is simply unattainable. It would require a total of 750 million tons of new mine capacity between 1978 and 1985. This is based on 150 million tons of new capacity due to depletion of existing Appalachian and Midwest mines plus 600 million tons of additional capacity over the current level of production. This would mean that in the seven years from 1979 to 1986, we would have to add 107 million tons of new capacity every year from now until 1985. This is about ten times the new capacity added each year during the past twenty-five years.

Anyone who has the least bit of knowledge about energy and the coal industry would immediately recognize that the idea of developing 100 new one-million ton mines each year for the next seven years is sheer fantacy.

The tragedy of being in this dream world is that if enough people believe that coal production can reach a level of 1.265 billion tons by 1985, they will also conclude that there will be no need to accelerate the nuclear program and they will also believe that imports of oil can be reduced. The horrible truth is that if coal production can only reach a level of 850 million tons by 1985, the shortage of 465 million tons under the Carter Plan will have to be made up with imported oil - and that difference is five million barrels per day.

Further, if our nuclear plant program is delayed even more, each plant not operating in the 1985-1990 period will result in the requirement for 10 million barrels per year of imported oil. If 100 nuclear plants are delayed, this will require an additional three million barrels per day of oil--all imported.

Since the Carter Plan projected that seven million barrels per day of imported oil will still be required by 1985 (about the same as 1975), we could be importing as much as 15 million barrels of oil per day by 1985 due to our inability to produce the required coal production and to construct enough nuclear plants. The two important questions we need to answer are where will this much imported oil come from and how can we afford to pay for it. As a matter of interest, if imported oil by 1985 costs $40 per barrel, the annual cost to the United States will be $220 billion per year. This compares with the 1977 imported oil cost of about $50 billion.

If we continue to impede the production of domestic oil and natural gas we will, in effect, be supporting the price of imported oil, particularly the OPEC price. The U.S. imports one-third of all the oil which is exported and if we were able to reduce our imports of oil we could affect the price. Is it possible that the U.S. is deliberately preventing or impeding the production of our domestic fuels in order to prop up the price of OPEC oil?

On December 4, 1978, a new study of "The United States Coal Industry: Problems and Prospects" was sent to the Members of the Permanent Subcommittee on Investigations by Senator Henry Jackson, the Chairman. This study was prepared by the Congressional Research Service of the Library of Congress. The study was prepared at the request of Senator Charles H. Percy, the ranking Minority member of the Subcommittee.

This study contains information on the current state of the coal industry including the coal resource base, trends in coal production and the demand for coal and coal mining technology. In addition, it addresses a number of complex issues including labor-management relations, Federal coal leasing policy, and Government regulation of the coal industry.

In Senator Percy's covering letter of December 4, 1978, transmitting this study to Senator Jackson, he pointed out that the coal industry in this nation faces a number of serious challenges. President Carter has singled out coal as an increasingly significant source in the years ahead. He has urged that coal production be doubled by the year 1985. Senator Percy's letter goes on to say "it will be an exceedingly difficult undertaking to achieve that goal. Over the last decade, a number of factors have caused productivity to decline markedly (to one-half the

1969 underground productivity level) and the price of coal to double. If these trends continue it may not be economically feasible to reach the President's goal."

"Another area where improvements must come involves government regulations of the coal industry. This report contains detailed descriptions of many of the environmental and health and safety laws which relate to the industry. Those laws have vitally important objectives which must not be compromised. However, it is clear that the regulatory process can be made less cumbersome to the industry without sacrificing its important goals. Regulations which have proven over time to be ineffective or overly cumbersome should be identified and either thrown out or rewritten. Exceedingly broad regulations should be tightened up to prevent different enforcement officers from giving widely varying interpretations of them. Efforts should be made to improve the calibre of enforcement officials, and to eliminate unnecessary delays, excessive paperwork and overlapping authority in the regulatory process."

"In sum, literally thousands of regulations affecting the coal industry have been issued in the last decade. Entire enforcement agencies have been assembled in the same short period. The time has come to subject those agencies and the regulations they enforce to an intensive review."

The study by the Environment and Natural Resources Policy Division of the Congressional Research Service of the Library of Congress is an extremely well prepared report. Excerpts from the study are quoted below:

"Despite the improving circumstances brought about by the passage of the National Energy Plan, coal's future is beset by forces which threaten to frustrate the Nation's objective of greatly increased coal use. Newly legislated regulatory requirements have added large costs to coal production and use. The regulatory processes themselves, often imperfectly implemented within agencies and poorly coordinated between agencies, have added and continue to add additional costs, have created extensive delays, and have introduced great uncertainty as to what will be required and when approvals will be given."

The Library of Congress study has these comments on synthetic oil and gas from coal. "For the past several years, there has been renewed interest in converting coal, which is so plentiful, into liquid and gaseous fuels which are not plentiful from domestic sources. The major products under consideration are solvent refined coal (SRC), oil from coal, synthetic natural gas, and medium- and low-Btu gases. The Department of Energy (and its predecessor agency, ERDA), many parts of the Congress and a number of private interests have been involved."

"Under the best of circumstances, none of these synthetic fuels could become significant commercial realities until the mid- or late-1980's, because of technological and regulatory uncertainties, plus long lead times. But of more fundamental concern is the economic question; all (except the low- and medium-Btu gases in tailored situations) appear unable, given current or foreseeably-applicable technologies to produce oils or gases at attractive prices. Instead, the proposals currently on the table rely on rolling in the high prices of synthetic fuels with the low prices of price-controlled existing supplies in order to come out with acceptable average fuel costs."

"Synfuels supporters grant currently unattractive economics, countering with arguments that ten or so years from now, with oil prices much higher than current levels (a debatable projection), synthetics will be able to compete economically. But this outlook is also questionable, according to the pessimists, because it does not take into account the impact of the then-higher energy prices on future capital costs of the synfuels plants."

"In sum, there is little chance of any significant market for coal for synfuels production for a decade, and a resonable chance only for a couple of demonstration plants by the year 2000. Thus, the coal industry is counting only on a demonstration plant market of perhaps five to six million tons per year."

On the subject of regulatory restraints on the coal industry the Library of Congress study had this to say:

"To the coal industry, this intricate, time consuming, expensive network of regulations means reduced growth potential, reduced flexibility for response to changing market conditions, and pressures for further concentration into fewer, larger companies within the industry. The regulatory network currently in place and in process of being put into place carries with it necessary new costs, additions to project lead times, and requirements for additional supervisory and managerial skills."

"Evidence presented in litigation, administration and Congressional hearings, and the press, show clearly that the several regulatory programs are neither optimally implemented nor optimally coordinated, thus adding potentially avoidable cost and scheduling penalties."

"Most additional cost imposed by regulatory requirements is reflected in coal price, both at the mine mouth and at the point of use. Coal is in competition with other fuels in all its markets; increases in the cost of using coal will reduce the incentive for increased coal uses and hence the use of coal."

According to information received from the Department of Energy on January 24, 1979, there has been a major revision downward in the projected coal production target for the year 1985. While the Carter National Energy Plan of April 1977 set a coal production goal of 1.265 billion tons by 1985, the new goal has been set at 900 million to 1 billion tons a year as a more likely figure for 1985. Energy Deputy Secretary John O'Leary told an Energy Department Conference during the week of January 15, 1979, of the new coal production forecast. At the same time, Mr. O'Leary pointed out that the 650 million tons produced during 1978 was about the same as production in 1918 and 1947 - which is on about a 30-year cycle.

We must recognize before it is too late that we must concentrate on the use of coal and nuclear and remove the obstacles which prevent these two important sources of energy from reaching their full potential and we must remove the price controls from oil and natural gas in order to allow our domestic resources to be developed and produced at their full potential. While we will continue to need imported oil for many years, we should do everything possible to minimize its use. Unfortunately, we are doing everything possible to prevent the domestic production of coal and oil and gas and to restrict the use of nuclear power. As a result, our imports of oil are increasing every day and in 1978 accounted for nearly 45 percent of our oil consumption and more than 27 percent of our energy. We are headed for self-destruction because we have failed to understand the complex relationship between energy and the economy and the catastrophic effects which imported oil is having on our ability to control our own destiny.

It is necessary that we do the following as quickly as possible if we are to remain a viable industrial nation:

1. Decontrol prices for new natural gas.
2. Decontrol prices for all crude oil and all petroleum products.
3. Amend the Clean Air Act to allow the burning to high-sulfur coal through use of intermittent control systems.
4. Remove obstacles to mining of coal through amendments to the Federal Mines Safety Act, Federal Coal Leasing Act and the Federal Surface Mining and Reclamation Act.
5. Remove obstacles to the construction and operation of nuclear power plants.
6. Pass legislation to provide for incentives for energy conservation such as a 50 percent tax credit during the first year for installation of coal-fired boilers and energy conservation equipment by industry and similar tax incentives to home owners for installing insulation, etc.
7. The Federal Government should finance synthetic fuel plants based on coal to produce liquid fuels and synthetic pipeline gas by establishing a tax on gasoline of ten cents per gallon. Such a tax would provide over $11 billion per year or enough funds each year to build 11 synthetic fuel plants each year from now until eternity. This is how South Africa finances its SASOL projects for converting coal to liquids and gases. Can we not be as smart as the South Africans?

8. The most important change we can make is to let the free marketplace determine prices for fuels and energy. This will allow the pricing mechanism to work and let consumers choose which fuels they wish to use and at the same time give producers the incentives necessary to increase the domestic production of fuels and energy.
9. Finally, contact your Senators and Representatives in Congress. Tell them by wires, phone calls, letters and in person how you feel about the energy situation. Remember, your future is at stake too!!

Summary

The United States of America has been for many years and still is the Number One industrial nation in the world. It reached this position because it has plentiful supplies of raw energy sources, other raw materials, skilled labor and management and the free enterprise system under which to operate.

Recent events have cast their shadows on the future role which the United States will play in the world and these events concern themselves with our ability to obtain sufficient energy and fuels to meet the demands of an expanding economy.

It is important that officials at all levels of government; executive, legislative and judicial branches, both federal and state, understand the consequences of their actions as they related to energy, the environment and to the future existence of the United States as an industrial nation. Everyone should realize that the only way we are going to solve our energy crisis is to allow the marketplace and the free enterprise system to work, unimpeded by the government. In becoming the world's industrial leader, the United States has also become the world's largest user of energy and accounts for about one-third of all the energy consumption in the world. We also produce over one-third of all the world's goods and services.

It is becoming increasingly important that we recognize that all of the future increase in the GNP in the United States for the next ten years or more is dependent upon imported oil. The bulk of this increased oil requirement will come from the Middle East and particularly from one country--Saudia Arabia.

Our domestic oil production and natural gas production have been declining at the rate of about five percent per year. Natural gas production in 1977 and 1978 stopped declining and this may be a signal that the marketplace is alive and well and operating at least at the intrastate level. Our coal production in 1978 was less than 1977. Nuclear power will also increase our supply of domestic energy in 1978 but there have been no new nuclear plants ordered during the past two to three years.

While we imported approximately 7.3 million barrels per day in 1976, we imported about 9 million barrels per day of oil in 1977 in order to have a real growth of five percent in the GNP. Likewise, the 1978 energy requirements required us to import nearly 9 million barrels per day of oil.

The costs to the U.S. for imported oil in 1977 and 1978 amounted to nearly 50 billion dollars per year based on an average delivered cost of $14 per barrel. We imported nearly 45 percent of our oil in 1977 and 1978.

As the production of natural gas and petroleum declines in the United States in the future, the possibility of producing substitute gaseous and liquid fuels from coal would seem to offer a solution to the shortage of convention fuels.

However, federal regulations are working against development of a commercial synthetic natural gas industry at a time when the nation needs more gas even at substantially higher prices. Pricing regulations and other controls have created disincentives and uncertainties which have discouraged investments in synthetic natural gas.

A total of 19 commercial-size high-Btu gas projects have been announced for the United States during the past few years.

On November 8, 1978, it was reported in the Wall Street Journal that "The only active commercial coal gasification project has been halted by American Natural Resources Co. and four gas-pipeline concerns. The suspension of design and engineering work on the $1.4 billion project in North Dakota followed the failure of attempts to settle a controversy over the method of financing the project which had delayed federal approval required before work could proceed."

"The Federal Energy Regulatory Commission, an independent agency within the Department of Energy, must authorize the interstate sale of gas produced at the plant."

"The American Natural Resources Co. had difficulty raising money for the mammoth project which would have produced 125 million cubic feet per day of pipeline quality high-Btu gas. Officials of the company had said that normal debt financing isn't possible because lenders aren't willing to put up the money for what is still an untried process in the United States. Yet outside financing is needed because the partners in the project can only afford to put up 25% of the cost of the plant themselves."

"Earlier in the project, the sponsors tried to get federal loan guarantees but this failed and then a plan was devised that would have guaranteed that in the event the project failed to be completed, lenders would be repaid with cash raised by increasing the monthly bills of millions of retail customers served by the sponsoring companies. The five companies involved--pipeline units of American Natural Gas Co., Peoples Gas Co., Transco CO., Columbia Gas System, Inc., and Tenneco, Inc.--serve between 12 and 14 million customers."

"The plan to backstep the project with consumer cash has been opposed by six states so far."

"One other complication which made necessary an approval of the project by January 1, 1979, was that Basin Electric Cooperative, which is building a separate electric generating plant at the North Dakota site, needed to know if the gasification plant would be hooked up to it or would provide its own power at the site."

"American Natural believes a link up with Basin is essential to the success of its project. With the two plants working together, there are great cost savings available as the coal gasification plant could receive electric power in exchange for excess coal."

"A spokesman for Combustion Engineering, Inc. which is providing engineering; procurement and construction services for the project, said it had advised suppliers to suspend any response to inquiries about meterials and equipment. Vendors were advised that the project is in a state of suspension." (2)

"A one-year delay could cost the group sponsoring the project formally called the Great Plains Gasification Associates, about $60 million in increased costs, American Natural said."

This project was originally announced in 1974 and at that time it was estimated that the cost of the first of four 250 million cubic feet per day unit would be $770 million. As time went on, the size of the project shrank to 125 million cubic feet per day and the cost ballooned to over $1.4 billion.

It should also be noted that when this project was first announced in 1974, the estimated cost of the gas was $4.00 per Mcf of 975 Btu/cf pipeline gas for 1980 production.

The latest cost estimate for the gas is that it will cost between $6.25 to $8.25 per Mcf with an approximate cost of $7.25 per Mcf at the gasification plant in 1983.

The FERC ruling on this project was to be handed down in January 1979, but it is not expected now until sometime this summer.

Robert D. Thorne, former Assistant Secretary for Energy Technology of the Department of Energy, resigned abruptly in December 1978 with the following observations about the DOE's change in direction with regard to synthetic fuels from coal. Mr. Thorne, according to the Business Week issue of January 22, 1979, sensed a shift at DOE particularly in the coal area, away from "near-term payoff type technologies" to an emphasis on longer range research and development. The same Business Week article, pointed out that changes in other areas attest to their uncertain progress. In coal research, the DOE had planned to build two demonstration facilities to prove out technology for converting coal to clean-burning fuel for power plants. An unconvinced Office of Management & Budget (OMB) has reportedly cut that number to one, despite considerable wasted effort on the discarded alternative. Plans to go forward with a demonstration plant for converting coal to high-Btu gas, a program in which the government has already spent $400 million, are also headed for delay, if not cancellation.

A Federal Power Commission's Natural Gas Survey Task Force pointed out in a 1976 report that "the technology is available to convert coal to SNG and with refinement may reduce the cost of gas in time. But despite these facts, there is no concerted national policy toward overcoming the major obstacles to substantial progress."

Factors deterring commercial development of coal gasification technology, cited in the report were:

1. Legislative and regulatory uncertainties, such as,
   a. gas price regulation,
   b. divestiture proposals (horizontal and vertical),
   c. accessibility of federal coal,
   d. ambiguous environmental regulations,
   e. uncertain fiscal policy.
2. Uncertainty created by timing delays associated with approval processes, environmental reviews and litigation.

The FPC Task Force draft study makes four recommendations:

1. FPC regulations and policies should be changed to provide incentives,
2. A "roll-in" concept should be used in price regulations for coal-based synthetic gas to be supplied over interstate systems,
3. Permit procedures should be expedited,
4. Congress should enact legislation that would allow only a limited period for governmental action on projects so that a final decision could be obtained without causing undue delays.

On the other hand, there are thirteen commercial "SNG from Petroleum" plants currently operating in the U.S. Total design capacity of all thirteen plants is 1,334.5 million cubic feet per day or about 0.5 trillion cubic feet per year. The aggregate investment cost for all plants was approximately $650 million. The feed stocks include naphtha, natural gas liquids, propane and butane. The cost of SNG gas from these plants is as high as $5 to $6 per Mcf. The investment cost is very low for this type of SNG plant and is on the order of $522 per million Btu per day of capacity at a 50% load factor. This can be compared with the investment cost of high-Btu pipeline-gas-from-coal plants which is now on the order of $12,444 per million Btu per day of capacity based on the Lurgi process and 90% load factor. There are two other SNG plants in the planning or construction stage.

The FEA forecast in 1977 that by 1985 the capacity of this type of SNG plant will total 1.0 trillion cubic feet per year up from the current capacity in November 1978 of 0.5 trillion cubic feet per year.

The natural gas supply in the U.S. will either decline by one-half in the next 25 years or remain at its present level depending upon whether or not there is deregulation of the price of natural gas at the wellhead. The American Gas Association's October 11, 1976, forecast of the "outlook for Natural Gas to the Year 2000" shows the following projections:

	Natural Gas Supply--10^{15} Btu		
	1975	1985	2000
Continued federal wellhead regulation	19.2	14.6	10.6
Deregulation	19.2	20.4	20.0

The total energy supply to the year 2000 would be as follows according to AGA:

	Energy Supply--10^{15} Btu		
U.S. Domestic	1975	1985	2000
Coal	13.3	17.0	30.0
Petroleum	19.7	29.3	27.5
Nuclear	1.7	11.8	46.1
Other	3.2	4.8	11.8
Dry Natural Gas & Supplements*....	20.2	25.3	27.5
Subtotal U.S. Domestic.......	58.1	88.2	142.9
Imports of Oil	13.0	11.8	7.1
Total Consumption	71.1	100.0	150.0

*Assumes removal of federal fuel price control

New supplies of natural gas are expected to come from a number of sources as follows:

	Supplemental Gas--10^{15} Btu		
New Supply	1975	1985	2000
Alaskan Gas......................	--	1.2	1.5
Canadian Imports	1.0	0.6	1.0
LNG Imports	--	3.0	3.5
Advanced Fracturing	--	0.1	1.5
Subtotal	1.0	4.9	7.5
Conversion			
Gas from Coal	--	0.4	2.5
SNG from Petroleum	--	0.4	0.5
Subtotal	--	0.8	3.0
Grand Total..................	1.0	5.7	10.5

Assuming removal of federal fixed price controls, the total supply of natural gas and supplementals would be as follows:

	Total Gas Supply - 10^{15}Btu	
	1985	2000
Dry Natural Gas	20.4	20.0
Supplementals	5.7	10.5
Total	26.1	30.5

Petroleum and natural gas supply over 75 percent of the total energy consumed in the United States and these two fossil fuels are expected to play a continuing role as major sources of energy for many years to come. Oil, which currently supplies 46 percent of our total energy has become an essential part of our industrial and transportation and electric power sectors and without sufficient supplies of oil our economy would first falter and then collapse. Because of restrictive laws and Federal government actions, the production of petroleum in the U.S. has peaked and has been declining since 1970 when it reached 11.3 million barrels per day.

It is anticipated that it will continue to decline at a rate of 5 percent per year. The crude oil from the Alaskan North Slope, added 1.2 million barrels per day starting in 1978, to our domestic production and thus reduced our requirements for imported crude oil and refined products. By 1985, this supply will account for 2.4 million barrels per day and will become of increasing importance as the world oil production begins to decline in the late 1980's or early 1990's.

As a result of the continuing increased demand for petroleum and a continuing decline in the domestic production of crude oil, it has been necessary to import crude oil and refined products in ever increasing amounts. In 1970, the U.S. imported 1.324 million barrels per day of crude oil and 2.095 million barrels per day of refined products, a total of 3.419 million barrels per day. At the same time, our consumption of oil was 14.697 million barrels per day. Our imports of oil were therefore 23.3 percent which amounted to 10 percent of our total energy consumption.

In 1976, oil consumption reached 17.3 million barrels per day, our domestic production declined to 10 million barrels per day and we had to import 7.3 million barrels per day at a cost of $35 billion this year to keep our economy operating. Imports in 1976 accounted for 42 percent of all of the oil we consumed and these imports represented 20 percent of our energy--double what it was in 1970.

In 1978, the United States imported about 8.5 to 9.0 million barrels per day of oil in the form of crude oil and refined products and the total supply including domestic production was about 19 million barrels per day. We import more energy in the form of oil than we produce in the form of coal.

It is forecast that by 1980, our oil comsumption will be 22 million barrels per day and that imports will be 12 million barrels per day. Imports would then account for 55 percent of our oil and over 30 percent of our energy.

Under these circumstances, the payments which the U.S. must make for imported oil first become burdensome and then intolerable. In the case of 12 million barrels per day of imports by 1980 and a price of $20 per barrel, the payments for imported oil will increase to nearly $90 billion per year by 1980. By comparison, the value of oil imports into the USA in 1972 was $4.5 billion. In the opinion of many persons, we will not be able to afford to import oil which costs us $90 billion per year. Our balance of trade deficit would be so high that devaluation of the dollar would have to be done on a weekly basis.

Going one step further to 1985, imports could reach a level of 15 million barrels per day and with an estimated price of $40 per barrel, the cost would be $220 billion. If we allow this to happen, then we deserve the fate which is in store for us.

Unfortunately, the Western World has concluded that imports of oil from the OPEC group will increase from 27 million barrels per day at present to as much as 37 million barrels per day 1985. This conclusion has been reached despite efforts to conserve oil and develop alternate energy sources. It seems that the easiest way out is to simply import and burn oil and try to forget about the inevitable consequences.

The choices we have to change the direction we are going are to increase the production of domestic oil and natural gas and to concentrate on the use of our coal and uranium to produce electric power. Any other choice is suicidal and will result in the total collapse of the United States of America as an industrial power.

However, in order to accomplish the above objectives, it is necessary that we understand what needs to be done. Even more important, we must realize the tragic consequences which are in store for us if we fail to make the move to use our own fuel resources.

Coal, which represents 90% of our total fuel resources, must be allowed to play its very important role in supplying fuel for electric power general as well as the other essential uses. We have to not only remove all the obstacles which now prevent the coal industry and all other fuel energy industries from producing at their maximum capabilities, but we must also prevent the Congress from breaking up these industries.

Let me first say that I believe we <u>should</u> produce 1 to 1.3 billion tons of coal by 1985 if we are to prevent this country from becoming overwhelmingly dependent upon insecure, high-priced supplies of foreign crude oil and refined products. Our increasing dependence upon the Middle East as the source of this crude oil should be of immediate and serious concern to all Americans.

While I believe that coal production at the 1 to 1.3 billion tons per year level by 1985 is an essential part of our energy supply program, I do not see any signs that those in charge of our energy program understand the magnitude of the task facing us in order to reach that coal production goal. Even worse, other Government Agencies such as FTC, EPA, and MESA have been acting and are continuing to act in a way which prevents the coal industry from achieving the 1985 goal.

In order to increase the U.S. coal production to the 1 to 1.3 billion tons per year level by 1985, we must not only determine what is needed directly in the form of capital, manpower, equipment and similar requirements, but we also need to determine what other actions need to be taken directly and indirectly and what restraints there are from an environmental, legislative, political and social standpoint. Having determined what these are, we then must study how these obstacles and restraints can be removed, how long it will take to remove them and what alternative approaches there are to solving the problems we perceive.

When we undertake to develop a new coal mine, it is common practice to set up a Critical Path Method of controlling the construction during the five to eight years of the development period. Since the number of "activities" amounts to from 500 to 1,000, it is necessary to set the program up on a computer and update it monthly.

I believe that we now need to expand the Critical Path Method of analysis to cover all the items which are directly and indirectly involved in the expansion of the coal industry to the 1 billion tons per year level. These include the usual ones such as capital, manpower, equipment, and, in addition, the environmental, legislative, political and social items. The latter have become as important as the other more common items.

The sulfur dioxide problem for example can be solved in a number of ways. The easiest way is by the use of intermittent control systems. However, EPA and Congress refuse to approve of this method. As a result, the requirements for low sulfur coal are increased by <u>ten</u> times over what they would be if intermittent control systems were allowed. The insistance upon the use of scrubbers is going to increase the cost of electricity by billions of dollars per year--all to be paid by the ultimate customers.

The confusion which has resulted from Federal and State Air Quality laws has prevented electric utilities from making future commitments for coal. As a result, coal companies are not making the commitments to new coal mines either. Many of the new mines which are listed as planned in current forecasts have already been delayed due to environmental suits and other restraints. Examples are the Kaiparowitz project in Utah and a number of other Wyoming coal projects.

There are similar problems which will prevent coal production from reaching 1 to 1.3 billion tons by 1985. These include the following:

1. Clean Air Act of 1970 and 1977
2. Federal Mine Safety Act of 1969

3. Federal Leasing Regulations
4. Transportation Facilities
5. Technical Manpower
6. Mine Labor
7. Equipment Availability
8. Federal Surface Mining and Reclamation Act

We need to set up a network or arrow diagram based on a Critical Path Analysis of all the factors and activities which relate to coal mining and determine in a quantitative way what is required and what specifically has to be done and by whom in order to achieve a coal production level of 1 billion tons per year.

We need to use the Critical Path Analysis in order to determine what work has to be done before other activities can be started and how all the activities related to each other.

The development of a network of activities or arrow diagram in which all activities have to be laid out end to end and which shows the relationship of each activity to all other activities would prove that EPA regulations with regard to sulfur dioxide will prevent new Eastern coal mines with high sulfur coal from ever being developed and will in time shut down all mines currently in production with high sulfur coal. Likewise, environmental suits and leasing delays will prevent new Western coal mines from ever being developed.

We also need to determine how many additional tons have to be mined due to the lower heating value of the Western coals. Coal with a heating value of 8,000 Btu per pound will require 50% more tons to give the same heating value as Eastern coal with a heating value of 12,000 Btu per pound. This means that we really will need 1.474 billion tons of coal by 1985 instead of 1.265 billion tons if one-third of the coal is to come from the Western states.

I believe it is possible to determine the answers to these questions but it will be necessary to analyze the problems systematically and on an interrelated basis.

As a result of the overkill provisions in the Clean Air Act, over fifty percent of the coal now being burned by electric utility power plants is non-complying coal due to its high sulfur content. It is obvious that the Clean Air Act has to be amended to allow the burning of high sulfur coal. Yet Congress refuses to acknowledge this logical solution and instead has passed even more restrictive legislation.

The Federal Coal Leasing Act of 1975 has been labeled a "procedural monstrosity" by the National Coal Association.

Faced with the above obstacles to mining enough coal for conventional uses, it is difficult to see how we can develop a synthetic fuel industry based on coal unless we make it more attractive from an economic standpoint.

Consumption of coal in a typical synthetic natural gas plant will amount to approximately 8 million tons per year per plant each with a capacity of 250 million CF/Day.

The February 1976 FEA forecase for coal-based SNG plants shows a total of 16 million tons per year of coal requirements by 1985, implying that 2 SNG coal based plants will be in operation and producing 0.16 x 10^{15} Btu per year. The AGA October 11, 1976 report in "Prospects to the Year 2000" shows the following forecast of the production of SNG from coal and petroleum:

	10^{15} Btu/Year	
	1985	2000
SNG-Coal Based	0.4	2.5
SNG from Petroleum......	0.4	0.5
Total..............	0.8	3.0

The AGA forecast for coal gasification in terms of numbers of plants and coal consumption is as follows: (Note: Assumes that each plant produces 0.08 TCF/Yr. and requires 8 million tons of coal per year.)

SNG-Coal Based	1985	2000
10^{15} Btu/Yr.	0.4	2.5
Number Plants	5	31
Million Tons Coal/Yr.	40	248

While on the subject of coal gasification, we should remember that low- and medium-Btu gas from coal should be considered as lower cost alternatives to pipeline quality high-Btu gas. For one thing, the capital investment for low-Btu gas plants is lower than the high-Btu gas plants as shown below:

Daily Btu Output	Number of Gasifier Vessels	Total Capital Investment Required	Investment $ Per Million Btu/Day-90% OF
Low-Btu Gas Plants (175 to 200 Btu/cf)*			
$2x10^9$	1	$ $6x10^6$	$3,333
$4x10^9$	2	$9.5x10^6$	2,639
$8x10^9$	4	$15x10^6$	2,083
$12x10^9$	6	$20x10^6$	1,852
$16x10^9$	8	$25x10^6$	1,736
$20x10^9$	10	$29x10^6$	1,611
$24x10^9$	12	$34x10^6$	1,574
Medium-Btu Gas Plant (350 Btu/cf)*			
$8x10^9$	4	$ $20x10^6$	$2,778
vs. High Btu Pipeline Gas Plant (Lurgi)			
$250x10^9$ (C.F. Braun - 1976)		$$1,070x10^6$	$4,771
$125x10^9$ (American Natural Gas Company - 1978)		$1,400x10^6$	$12,444

*Data from Holly, Kenney, Schott, Inc. of Pittsburgh, PA. based on use of Woodall-Duckham coal gasification process.

In a survey prepared by Stone & Webster Management Consultants for the Edison Electric Institute, of 142 companies in the 15 most energy intensive industries surveyed, 114 companies indicated they expect a shortfall of certain types of fossil energy. When these respondents were asked what fuels they expected to be in short supply, 126 responses were made: 113 anticipated natural gas shortages, 11 expected oil supply problems, and 2 questioned the long-term availability of electric power. Clearly, the natural gas industry has a problem on trying to hold on to their existing industrial customers who have been curtailed at ever increasing amounts during the past five years. All signs point to a continuation of these curtailments under the present regulatory climate.

The natural gas industry has found itself in a situation in which its industrial customers are running away from it faster than the available supply of natural gas is declining. Under these circumstances, there will be an excess supply of natural gas--not a shortage. Therefore, closer ties between the natural gas industry and its industrial customers must be set up and maintained.

As a result of the federal government's actions in trying to force industrial plant and electric utility plants away from natural gas, the U.S. found itself in late 1978 with large surpluses of natural gas. The Department of Energy also belatedly came to the startling realization that the natural gas industry couldn't operate if its only customers were the residential and commercial markets which are seasonal in nature--probably only five months out of the year at most.

Secretary Schlesinger stepped into the breach and announced in December of 1978 that the Department of Energy wanted industries and utilities that now burn oil to switch in the short term (three to five years) to natural gas, not coal as called for by the National Energy Act. Such a revised policy, would reduce the U.S. dependence on imported oil and strengthen the dollar.

On January 9, 1979, Secretary Schlesinger in a talk to the National Association of Petroleum Investment Analysts Group of New York in New York City, again called on industry to switch back to natural gas from oil. In a major shift in the Carter Administration policy, Secretary Schlesinger said that the United States would emphasize increased industrial consumption of natural gas instead of coal to reduce oil imports.

Mr. Schlesinger was quoted in the New York Times (January 10, 1979) as saying, "Although the Administration remains committed to the use of coal instead of oil or gas in new boiler facilities over the longer run, over the course of at least the next several years, existing industrial and utility facilities will be provided every encouragement to burn gas instead of oil"

Secretary Schlesinger estimated that because of Government policies as well as the effects of a serious shortage during the winter of 1976-77, some three trillion cubic feet of natural gas per year that could otherwise be used was not being consumed.

The U.S. currently produces about 19 trillion cubic feet per year and imports an additional one trillion cubic feet from Canada.

The main reason for the surplus according to Secretary Schlesinger is the fact that high prices have stimulated a surprising amount of new production of natural gas. (This may be surprising to Secretary Schlesinger and other government bureaucrats, but it certainly isn't surprising to those of us in the private sector who have believed for many years that all that is needed to solve the energy crisis is a free marketplace and removal of price controls from energy supplies).

As a result of this sudden about-face by the Carter Administration and their new found religion in switching to natural gas, it shouldn't surprise anyone to find there will be little incentive for anyone to keep working very hard on coal gasification whether it be for low-Btu, high-Btu or medium-Btu gas or any combination of them.

Likewise, the coal industry which has been sent reeling from one blow after another from the federal government including mine safety, surface mining regulations, leasing regulations and hazardous materials regulations, now finds it has been told to wait a few more years before it will really be needed.

Consolidated Edison Co. announced on January 12, 1979 that it will seek federal approval to replace up to 10 million barrels of imported oil per year with domestic natural gas as a fuel in its electric power and steam plants.

The announcement as reported in the Wall Street Journal was prompted by a statement made earlier in the week by Energy Secretary James Schlesinger recommending such substitutions.

While the Carter Administration has long discouraged the use of natural gas to fuel utility and industrial plants, it recently reversed this position because a short-term surplus of natural gas has developed and they are now pleading with electric utilities and industrial plants to convert back to natural gas.

The New York Times reported on January 19, 1979 that "about $1 billion of the national budget savings that President Carter has ordered to keep next year's federal deficit under $30 billion has come out of the energy budget according to government and industry officials."

In the original appropriation budget request by the Department of Energy the total was $9.1 billion for the 1980 fiscal year starting October 1, 1979. The actual spending request for the 1980 budget was set initially by the DOE at $8.2 billion. However, the Office of Management and Budget slashed the appropriations request by 23 percent to $7 billion and the projected spending figure to $ 6.9 billion.

Secretary Schlesinger's appeals to OMB to increase both the appropriations and spendings were partially successful since the OMB has finally approved a figure of about $8 billion for both appropriations and spending for fiscal 1980.

One of the casualties of the budget cutting by OMB was reported to be funds for the commercial application of synthetic gas from coal. The original spending request was $224 million and the OMB slashed this to zero. Secretary Schlesinger asked that $98 million be restored. The final outcome is not known as of January 20, 1979.

The biggest single action that will help the entire supply-demand relationship in natural gas is to decontrol the wellhead price of all new gas. Until this is done, we can expect to see irrational actions by all involved.

Nuclear power, which now accounts for about 3 percent of our total energy consumption, (about 12% of the total electric generation) has to be allowed to grow rapidly in order for it to provide its share of energy which will of course be based on domestic uranium reserves. If uranium power is to account for nearly 30 percent of our total energy consumption by the year 2000 (compared with 3 percent now), the shackles and obstacles must be removed as rapidly as possible.

As of November 29, 1978, there were 72 operating nuclear power reactors in the United States with generating capacity of 52,273 megawatts (MWe). The total number of plants committed is 203 with a total capacity of 197,918 MWe.

Do not be mislead by politicians who announce that nuclear power should be used only as a last resort. The fact is that without nuclear power neither the United States nor any of the industrial nations of the world can long exist as industrial nations without relying on nuclear power for their principal source of energy. The sooner our leaders recognize this fact, the faster we can start solving our energy problems.

If we do not use our own coal and uranium to their maximum potential, then we do so at our peril because we cannot exist as as industrial nation. There simply is not enough petroleum and natural gas to give us the equivalent energy. Failure to recognize this vital fact will lead us to our inevitable doom. An enlightened leader can prevent us from becoming a satellite of the Middle East or a fourth rate power.

Despite the fact that coal reserves in the United States are sufficient to last for hundreds of years, coal consumption in the U.S. now accounts for only 18 percent of our total energy requirements. Even more surprising is that the recent U.S. Bureau of Mines study of "Energy Through the Year 2000" predicts that coal will account for only 21 percent of the total U.S. energy consumption by the year 2000 compared with 18 percent in 1975. By the year 2000, according to the Bureau of Mines, oil and gas are expected to supply only 44 percent but nuclear power is expected to account for 28 percent of the total U.S. energy consumption by the end of this century up from about 2 percent in 1975.

But a "worse case" condition could result in far greater imports of oil than the Bureau of Mines is forecasting. Unfortunately, coal will not be able to fill its relatively minor role in the U.S. energy mix unless we quickly come to our senses and remove all the obstacles which prevent coal from realizing its full potential.

We need to not only remove all the obstacles which now prevent the coal industry and all other fuel and energy industries from producing at their maximum capabilities, but we also must prevent Congress from breaking up the fuel and energy industries. The Congress is now considering bills which would split the oil industry into four major components. If this is done, petroleum products will cost the ultimate consumer more than they now cost him due to the inefficiencies which will result from the fragmentation of the oil industry. Other agencies in the government such as the Federal Trade Commission are busily engaged in trying to split off coal and uranium companies from their oil and mining parent companies. This will be the height of folly in a world where folly has become fashionable for government planners and agencies.

In the case of the coal industry, only the massive infusions of capital from their parent companies has kept many coal companies financially viable and allowed coal capacity to be expanded as the result of the expenditure of billions of dollars in the past ten years. If it were not for this additional new capacity, the level of coal production would be down around 400 to 500 million tons per year instead of 670 million.

Lest we forget, the only way this country is going to solve its energy problems is to let the marketplace and the free enterprise system work. Get the government off our backs and let people and companies who know how to find and produce oil and gas do so and let the people and companies who know how to mine coal and uranium do so.

We should always remember that the government has not produced one barrel of oil, not one cubic foot of natural gas, not one pound of uranium oxide, and not one ton of coal. All the government has ever done is prevent most of our fuels from being produced and then prevented those fuels which were produced from being burned.

What this country needs and needs now is a real national energy policy. Congress and the Administration must recognize this and prepare such a policy before it is too late.

When the lights go out for the last time and the factories and plants grind to a halt, it will be too late to realize that the government has destroyed the last free place on this earth.

Details on Coal

The United States is fortunate in having one of the world's largest reserves of coal. Total measured and indicated reserves of coal in beds over 28 inches thick and under less than 1,000 feet of overburden totaled 434 billion tons as of January 1, 1974. Of these reserves, 297 billion tons were considered underground reserves and 137 billion tons capable of being mined by surface mining methods.

Geographically, 47 percent of these reserves occur east of the Mississippi River with the remaining 53 percent in the Western States and Alaska.

Three-fourths of the strippable coal and one-half of the coal which can be mined by underground methods are west of the Mississippi River.

Since the recoverable reserve figure is not the most important number, the above reserve tonnages have to be divided by two based on 50% recovery in order to show recoverable reserves. After taking into account the deductions from reserves due to losses in mining, the total amount of recoverable reserves amounts to 148.5 billion tons of underground coal and 68.5 billion tons of surface coal reserves or a total of 217 billion tons.

Coal should and could be a partial answer to our energy supply problems for the near term and intermediate term and provide as much as 20 percent or more for the next 25 years, if we only have enough sense in this country to recognize what we have to do to make coal production possible. Actually the combination of coal and nuclear power could give the United States a reasonable assurance of becoming relatively independent in terms of imported energy supplies.

I say "relatively" because we should recognize that we will always have to import some oil and natural gas. As mentioned before, we now rely on oil and gas for 76 percent of our fuel and energy.

We will continue to rely on oil and gas for most of our fuels and energy for many years to come and a large proportion of this will have to be imported.

To further amplify on just one of the obstacles to increase the U.S. coal production, the average underground productivity of U.S. coal mines increased from 10.64 tons per manday in 1960 to a high of 15.61 tons in 1969 and has been dropping steadily ever since to a level of 9.50 tons per manday in 1975 and down to 8.0 tons per manday in 1978. It should be noted that the Federal Coal Mine Health and Safety Act was enacted in 1969 and it is no coincidence that coal productivity has been declining ever since. As a result of the impact of the Mine Safety and labor unrest on underground mining productivity, the coal industry has, in effect, lost 50% of its deep mine capacity during the period 1970-1978. This has had the effect of eliminating over 200 million tons per year of productive capacity. We, therefore, have to develop 200 new coal mines each with a capacity of 1 million tons per year at a cost of $30 million to $50 million per mine capacity back to what it was before the Mine Safety Act was enacted in 1969!!

So it really doesn't make any difference whether we have 300 years of reserves of coal or 3000 years of reserves. Our leaders in Congress have made it impossible to mine coal and they have made it illegal to burn half of what is being mined.

The reasons for our inability to expand coal production are very simple and should be understood by everyone.

First, the only new coal mines which are going to be developed will have to be financed on the basis of take-or-pay contracts with prices sufficiently high to attract the capital needed for the investment. Profits as high as $10 per ton are required to finance new deep coal mines today.

Second, the take-or-pay contracts which are required to finance these mines have to be for long enough periods to amorize the investment in the mine, so 20-year or longer contracts have to be entered into.

Third, these new mines are going to take six to eight years to develop in the case of the underground mines and three to five years in the case of surface mines.

In recent years, Congress has entertained the idea of passing laws prohibiting or severely limiting the surface mining of coal. If a total prohibition were to be put into law, it would eliminate 32 percent of all the coal reserves in the U.S.

The Clean Air Act has had the effect of prohibiting the burning and thus eventually the mining of much of the underground coal which has over certain sulfur levels.

Only 11 percent of the eastern coal reserves contain 0.7 percent or less sulfur. Most of this coal is low-volatile metallurgical coal, unsuitable for burning in electric power plants and, in any event, more valuable for the production of coke required for steelmaking.

If as much as 5% of the eastern coal reserves are available for use as low sulfur fuel for utilities and if this is all that can be counted on for power generation due to western coal leasing problems and low heating value to sulfur content ratios, this could mean that only 5% of 102 billion tons or 5 billion tons or enough to last eight years could be available for mining. Let us all recognize that such a drastic reduction in our reserves is not going to occur unless everyone in Washington has taken leave of his senses. But it shows what little coal we have left if we continue our present march towards self destruction by first refusing to allow strip coal to be mined, then refusing to allow Federal coal to be leased, then preventing it from being mined and then refusing to allow high-sulfur coal to be burned. Unless these current and planned and proposed restrictions are removed, we will not be able to survive as an industrial nation.

In our opinion, it will be impossible to expand the U.S. coal industry to a level of 1 to 1.3 billion tons by 1985. This fact is slowly being recognized by our leaders in Washington and you will soon start to see lower estimates of coal production being forecast for 1985. For example, figures of 1,000 million tons, including 100 million tons for export by 1985, are now being circulated. But unless an authority begins to understand what even this lower level of production means in terms of the job to be done, even this lower forecast will not be attained. Let us give one example of the magnitude of the job we have to do. If the present coal production level of 700 million tons is to be increased to 1,000 million tons by 1985, we would have to increase the mine capacity by 300 million tons plus the mine capacity which will be depleted at the rate of about 3 percent for eastern coal capacity a year or 15 million tons per year which is 100 million tons of capacity in seven years. The total additional new capacity is, therefore, 400 million tons. If we assume that 300 million tons will be western coal, this will require 60 new five million tons per mines in the Western States. The balance of 100 million tons per year could be obtained by developing 40 new two million tons per year underground mines and 10 new two million tons per year surface mines in the East. This schedule which calls for constructing 110 new large mines in the next seven years, to 1985, needs to be compared with the number of new large coal mines which have started producing coal in the past seven years--since 1972.

According to the 1978 Keystone Coal Industry Manual, there are only 44 U.S. coal mines which produced two million tons per year or more in 1977. Of these, only six mines produced more than five million tons per year. Only 13 of these mines started production in 1972 or later and only four of the mines produced over five million tons per year each in 1977.

If we exerted a superhuman effort and if we removed all the roadblocks and obstacles to developing all the new coal mines which we need, we would probably still fall short of this revised and lower forecast 1,000 million tons per year by 1985. Since we see no hope that anyone in Washington either understands the problem or in fact seems to care, we believe it will be impossible to attain a level of coal production of 1,000 million tons by 1985.

If we are able to achieve a 2 to 3 percent new increase per year from 1979 through 1985, we would reach a production of 774 to 853 million tons by 1985. It is interesting to note that the average increase in coal production during the period 1960-1978 was 2.6 percent per year. The actual increase for 1976 was 2.6 percent over the 1975 production, and the 1978 production was lower than 1975, due to the miners strike in 1978.

We therefore believe that because of the wishful thinking in Washington and the lack of understanding of the magnitude of the problem, we will probably be producing only 800 to 850 million tons of coal per year by 1985. The only thing which will change this is a sincere recognition by our leaders that they--Congress and Administration--are the reason why the coal industry cannot produce as much coal as the country needs in order to survive. And having recognized this fact, that they--Congress and Administration--have to pass the necessary legislation which will allow the coal industry to do the job it is really capable of.

ANNUAL PRODUCTION - COAL - U.S.A

Million Tons

	3% Annual Increase	2% Annual Increase	Actual	
1975	635	635	648	1975
76	654	648	679	76
77	674	661	689	77
78	694	674	646	78
79	715	687	713 NCA Estimate	79
80	736	701		
81	758	715		
82	781	729		
83	804	744		
84	829	759		
85	853	774		

The question I ask is: "Can even this be done?" I would answer that it is possible if everyone really wanted to. But first some long term compromise on non-health-related air pollution standards would be necessary. Few investors will risk capital for mine development when pollution requirements could make coal mines environmentally obsolete years before their normal payout period.

As we can see from the recoverable reserve figures, we have many years supply of coal if we are allowed to mine it. But unless a commitment is made to coal by our Government which will remove the restrictions already in place, this coal will not be mined regardless of how many years of reserves there are.

In order to fully comprehend the serious nature of our energy situation, let's look at where we get our fuels and energy now and how we expect to get them 10 years and 25 years from now.

The United States consumed a total of 71 quadrillion Btu's in 1975 or 30 percent of the total world's energy consumption.

U.S. Consumption of Fuels and Energy in 1975

Fuel	Quantity	10^{15} Btu's	%
Bituminous Coal & Lignite	562 Million Sh. Tons	13.266	18.67
Anthracite	5 Million Sh. Tons	0.128	.18
Petroleum Products			
From Crude Oil	4.5 Billion Barrels	26.001	36.58
From Other Sources	.7 Billion Barrels	4.200	5.91
Natural Gas, Dry	19.7 Trillion Cu. Ft.	20.173	28.38
Natural Gas, Liquids	594 Million Barrels	2.500	3.52
Electricity, Water Power	304×10^9 kWh	3.158	4.44
Electricity, Nuclear Pow.	155×10^9 kWh	1.652	2.32
Grand Total		71.078	100.00%

The following tables show the U.S. 1977 Consumption; Production and Imports of Energy:

U.S. Consumption of Fuels and Energy in 1977

Fuel	10^{15} Btu
Coal	14.133
Natural Gas	19.931
Petroleum	36.947
Hydroelectric Power	2.511
Nuclear Electric Power	2.674
Geothermal and Other	0.103
Total	76.299

U.S. Production of Energy in 1977

Coal	15.926
Crude Oil	17.315
NGPL	2.323
Natural Gas	19.566
Hydroelectric Power	2.331
Nuclear Electric Power	2.674
Geothermal and Other	0.088
Total	60.223

U.S. Net Imports (Exports) of Energy in 1977

Coal	(1.417)
Crude Oil	13.764
Refined Petroleum Products	4.282
Natural Gas	.975
Electricity	.180
Coke	(.015)
Total	17.769

The National Coal Association Economics Committee issued forecasts in December, 1978, for U.S. coal consumption and production for 1979 compared with the 1975-78 period. This forecast and comparison is shown on the following page.

Also following, is the most recent DOE forecast for coal consumption and production which was issued February, 1979, and is more optimistic than the NCA forecast, particularly on the export prediction.

NCA Forecast

Millions of Short Tons

	Actual 1975	Actual 1976	Actual 1977	Actual 1978	Forecast 1979
Electric Utilities	403.2	445.8	474.8	481	510
Coking Coal	83.3	84.3	77.4	71	75
General Industry	58.8	60.5	60.4	61	65
Retail	7.3	6.9	7.0	7	7
Total Domestic	552.6	597.5	619.6	620	657
Canada	16.7	16.5	17.2	15	16
Overseas	49.0	42.9	36.5	25	31
Total Exports	65.7	59.4	53.7	40	47
Grand Total Consumption	618.3	656.9	673.3	660	704
Production	648.4	678.7	688.6	646	713

DOE Forecast

U.S. Coal (Million of Tons)

Consumption By Sector	1977[1]	1978[2]	1979[3]	1980[3]	% Change 78/77	% Change 79/78
Electric Utilities	475	483	541	577	+ 2%	+12%
Coking	78	72	78	80	- 8	+ 8
Industrial	67	66	71	73	- 1	+ 8
Total Domestic	619	622	689	729	+ 1	+11
Export	54	40	60	63	-26	+50
Total	673	662	744	792	- 2	+12
Production	689	654	754	780	- 5%	+15%

[1] = actual
[2] = estimated
[3] = forecast

USBM FORECAST OF GROSS ENERGY CONSUMPTION TO THE YEAR 2000 (1)

	1974		1980	
	Trillion Btu	% Total Gross	Trillion Btu	% Total Gross
Coal	13,169	18.0	17,150	19.7
Petroleum	33,490	45.8	41,040	47.1
Natural Gas	22,237	30.4	20,600	23.6
Oil Shale	-	-	-	-
Nuclear Power	1,173	1.6	4,550	5.2
Hydropower & Geothermal	3,052	4.2	3,800	4.4
Total Gross Energy Input	73,121	100.0	87,140	100.0

	1985		2000	
	Trillion Btu	% Total Gross	Trillion Btu	% Total Gross
Coal	21,250	20.6	34,750	21.3
Petroleum	45,630	44.1	51,200	31.3
Natural Gas	20,100	19.4	19,600	12.0
Oil Shale	870	0.8	5,730	3.5
Nuclear Power	11,840	11.4	46,080	28.2
Hydropower & Geothermal	3,850	3.7	6,070	3.7
Total Gross Energy Input	103,540	100.0	163,430	100.0

(1) "United States Energy Through the Year 2000", by USBM, December, 1975

PRESIDENT CARTER'S ENERGY PLAN - APRIL 1977

Millions of Barrels of Oil Equivalent Per Day

	1976	1985 (With Carter Plan)
Supply	37.0	46.4
Domestic		
Crude Oil	9.7	10.6
Natural Gas	9.5	8.8
Coal	7.9	14.5
Nuclear	1.0	3.8
Other	1.5	1.7
Refinery Gain	.4	.6
TOTAL	30.0	40.0
IMPORTS/(EXPORTS)		
Oil	7.3	7.0
Natural Gas	.5	.6
Coal	(.8)	(1.2)
TOTAL	7.0	6.4

In Quadrillion (10^{15}) BTU PER YEAR

	1976	1985 (With Carter Plan)
Supply	74.0	92.8
Domestic		
Crude Oil	19.4	21.2
Natural Gas	19.0	17.6
Coal	15.8	29.0
Nuclear	2.0	7.6
Other	3.0	3.4
Refinery Gain	.8	1.2
TOTAL	60.0	80.0
IMPORTS/(EXPORTS)		
Oil	14.6	14.0
Natural Gas	1.0	1.2
Coal	(1.6)	(2.4)
TOTAL	14.0	12.8

TABLE III

PRESIDENT CARTER'S ENERGY PLAN - APRIL 1977

DOMESTIC COAL PRODUCTION
(Millions of Short Tons)

Region	1975 Production	1985 President Carter's Program	Increase Over 1975
Appalachia	396	627	231
Midwest	151	221	70
West	101	417	316
National	648	1,265	617

Utility Coal Consumption

Region	1975 Production	1985 President Carter's Program
East	186	327
Midwest	173	246
West	45	206
Total	404	779

TABLE IV

PRESIDENT CARTER'S ENERGY PLAN - APRIL 1977

DOMESTIC COAL CONSUMPTION
(Millions of Short Tons)

Sector	Actual 1975	Carter Plan 1985
Electric Utility	404	779
Industrial	63	278
Metallurgical	83	105
Synthetics	-	12
Other	6	1
Exports	65	90
Stock Changes	27	-
Total	648	1,265

DOE ENERGY INFORMATION ADMINISTRATION

FORECAST OF COAL DEMAND AND SUPPLY - MAY 8, 1978

U.S. COAL[1] DEMAND, SUPPLY, AND EXPORTS

(Millions of Tons)

		U.S. Total Demand	U.S. Domestic Supply	Exports
Actual 1975		556	[2]648	66
EIA Series 1985				
B-High	Low	992	1,065	74
C-Medium	Medium	961	1,034	74
D-Low	High	921	994	74
EIA Series 1990				
B-High	Low	1,224	1,304	81
C-Medium	Medium	1,177	1,257	81
D-Low	High	1,065	1,145	81
		Average Annual Growth (percent)		
1960-1975 (Actual)		2.6	3.0	4.0
1975-1985 (Projected)				
B-High	Low	6.0	5.1	1.2
C-Medium	Medium	5.6	4.8	1.2
D-Low	High	5.2	4.4	1.2
1985-1990 (Projected)				
B-High	Low	4.3	4.1	1.8
C-Medium	Medium	4.2	4.0	1.8
D-Low	High	2.9	2.9	1.8

1 Bituminous coal and lignite
2 Including production for stock-building
Note: Data may not add to total supply due to rounding.

POTENTIAL SUPPLEMENTAL SOURCES GAS - USA (1)

TRILLION CUBIC FEET PER YEAR

SOURCE	1977 ACTUAL	1980	1985	1990	1995	2000
Canadian Imports	1.0	1.4	1.4	1.1	1.0	0.8
SNG[1]	0.3	0.5	0.9	0.9	0.9	0.9
LNG Imports[2]	0.01	0.6	1.6	2.4	3.0	3.0
Mexican Imports	-	0.4	0.7	1.0	1.0	1.0
Alaskan Gas						
Southern[3]	-	-	0.1	0.2	0.3	0.6
North Slope[4]	-	-	0.7	1.4	2.2	3.0
Coal Gasification[5]	-	-	0.2	1.2	2.4	4.0
New Technologies[6]	-	-	0.1	0.5	1.0	1.5
TOTAL	1.31	2.9	5.7	8.7	11.8	14.8
US 48 STATES WITH DECONTROL	20.0	19.6	20.0	20.1	20.0	20.0
TOTAL SUPPLY	21.31	22.5	25.7	28.8	31.8	34.8

[1]Estimate for 1980 includes plants in operation. Estimates for 1985 and beyond includes plants which are approved, planned and suspended. All estimates assume year-round operation.
[2]Estimates for 1980 and 1985 are based on only announced projects.
[3]Southern Alaska includes onshore and offshore production south of Artice Circle.
[4]Assumes second major gas transportation system in operation by the early 1990s.
[5]High Btu gas only. Assumes suitable financing assistance (such as loan guarantees) for first few projects.
[6]Degasification of coal, gas from Devonian shale, gas from tight formations, gas from geopressured zones, gas from biomass and gas from in-situ coal gasification, etc.

(1) Data from American Gas Association Forecast in Gas Supply Review May 1978, Vol. 6 No. 8.

LITERATURE CITED

1. The Wall Street Journal, January 15, 1979. Reprinted with by permission of The Wall Street Journal, © Dow Jones & Company, Inc. 1979. All rights reserved.

2. The Wall Street Journal, November 8, 1978. Reprinted by permission of The Wall Street Journal, © Dow Jones & Company, Inc. 1978. All rights reserved.

L. PETRAKIS: I was somewhat confused. Yesterday we heard a great deal of discussion about the role that the government should be playing in liquefaction. This morning we heard an impassioned plea for the government to step aside. Are the two positions reconcilable?

G. GAMBS: My basic philosophy is that the government shouldn't have any role at all on the producing side. I will admit that if you wanted to make a case that we are better off taking government money, which basically is tax money, and putting it into synthetic fuels because that's the only way it is really ever going to come about, then I would rather do that than I would pay money forever to the Middle East. So I think that in a way I have schizophrenia about the government position. But I agree with all of you--there is no way this thing is going to fly without some government sponsorship, and you have to rationalize that on the basis that it's better to keep that money here in this country than it is to send it overseas to Saudi Arabia, Iran, or Kuwait, or wherever you want to send it.

RECEIVED September 10, 1979.

10

Major Technical Issues Facing Synthetic Pipeline Gas

L. E. SWABB and H. M. SIEGEL

Exxon Research and Engineering Company, P.O. Box 101, Florham Park, NJ 07932

Exxon has been doing research on coal gasification for over ten years. The early part of this work was aimed at making hydrogen for coal liquefaction, but in more recent years we have been working on synthetic pipeline gas, SNG, and intermediate Btu gas, IBG. In our early work, we identified what we believed to be an improved thermal process for coal gasification. We also began experimenting with catalytic gasification. By 1975, we concluded that the catalytic approach would be more promising on a long-term basis, and we shifted our work from thermal to catalytic gasification. Since mid-1976, the U.S. Department of Energy has been funding a substantial portion of our work on catalytic gasification for SNG.

As part of our total gasification program, we have made numerous design and cost studies to evaluate our process ideas as well as process systems being pursued by others. The comments that I will make today are based on the understanding of gasification systems that we have developed from this work.

Figure 1 shows what we believe are the main technical issues facing synthetic pipeline gas.

I will discuss the first five of these areas. The last area, on commercialization, will be covered in two other papers later this morning.

Figure 2 deals with potential gasification feedstocks in the contiguous 48 United States. As shown, coal, at about 5000 quads, is by far the largest recoverable fossil fuel. Peat is next at about 750 quads, and then oil shale at about 500 quads. And finally, we have also shown crude oil at about 170 quads to add perspective to the reserve estimates. These estimates, including crude oil, were published by the Institute of Gas Technology (IGT) in late 1977.

0-8412-0516-7/79/47-110-171$05.00/0

- What fossil fuel resources appear to be the most promising gasification feedstocks?
- How can new gasification processes reduce SNG cost...and how much reduction can be expected?
- What are the main challenges in planning the development program for a new gasification process?
- What is the outlook for the national R&D effort on new SNG processes?
- What is the impact of environmental considerations?

- What is the outlook for commercialization?

Figure 1. Main technical issues facing synthetic pipeline gas

COAL		QUADS (10^{15} BTU)*
SUBBITUMINOUS		3,100
BITUMINOUS		1,700
LIGNITE		200
ANTHRACITE		100
	TOTAL	5,100
PEAT		750
OIL SHALE		500
CRUDE OIL		170

* Two quads/year = one million B/D oil equivalent

Source: D.V. Punwani, et al (IGT), "SNG production from peat," December 1977

Figure 2. Recoverable fossil fuels in contiguous 48 states of the U.S.

As you know, most of the gasification work to date has been on coal. Coal is most abundant, and would also appear to be the most economical feedstock; that is, to produce the lowest cost SNG.

Oil shales, in both the western and eastern United States, offer another potential resource for SNG production. IGT has been doing research for several years on the hydrogasification of oil shale. They have shown that high recovery of the organic carbon in the shale can be obtained as gas and liquid products. From a cost standpoint, the main challenge is how to minimize and overcome the major cost of mining, crushing, feeding, processing, and finally discharging the very large shale volumes that must be handled per unit of gas product. The shale volumes are very large because the shale contains only 10-15% organic material to begin with.

Turning to peat, peat is the first product in nature's coalification process. About half of the total U.S. peat reserves are in Alaska, and the other half, shown here, is in the northern and eastern U.S. Gasification research on peat is relatively new. Overall, the material is highly reactive and can produce respectable yields of gas and liquids on a dry peat basis. But therein lies a key problem--how to get peat on a dry basis. Peat is about 90% water as mined. The challenge is how to remove this water without an overwhelming cost. It would appear at this stage that more research, design, and cost studies would need to be made before the practicality and competitiveness of peat gasification can be better assessed.

And now I would like to move on to the second issue: "How can new gasification processes reduce SNG cost...and how much cost reduction can be expected?" Figure 3 introduces this issue.

Commercially demonstrated gasification technology is available today to produce intermediate Btu gas from coal. I am referring to the Lurgi, Koppers-Totzek, and Winkler processes. Each of these processes has the potential to also produce SNG by the addition of shift and methanation steps downstream of the gasification system. As of 1977, Lurgi and Koppers-Totzek each had about 15 plants, and Winkler had 3 plants, operating in other countries to produce low and intermediate Btu gases.

Over the past seven years, a number of groups in the U.S. have announced plans for Lurgi coal gasification commercial projects to produce SNG. However, none of these projects has reached the construction stage. The main reasons for the delays have included problems with government approvals and regulations. Difficulties with environmental clearances, the cost and pricing of the gas, and financing arrangements. The technology has not been a limiting factor, and new technology now under development will not overcome these barriers.

And now, let's turn to the new technology, summarized in Figure 4. The new coal gasification processes now being developed for SNG have two main objectives: (1) to reduce cost, and (2) to process a wider range of coal types and coal particle sizes. Regarding cost reduction, we have listed on the slide the process improvement goals that we believe are likely to be the most fruitful in achieving lower SNG costs in new or improved processes.

The first goal is to reduce the required heat input to the gasifier. This can be done by producing more methane directly in the gasifier and less methane by downstream methanation. Lower gasification temperatures would also help. The next goal is to accomplish the heat input without using pure oxygen. Two possibilities for achieving this include the circulation of hot solids and the addition of a separate heat-liberating chemical reaction to the gasifier. I will come back to these first two goals later on.

Additional goals are to reduce equipment multiplicity; reduce the number, complexity, and size of individual process steps; improve heat recovery and utilization; and finally, to improve the operability and reliability of the overall plant system.

The challenge in developing lower cost SNG gasification processes is: (1) to combine as many of these items as possible into each new process; and (2) to accomplish this without adding so much additional cost in other parts of the processes so as to wipe out the savings.

In trying to achieve these improvements, a wide variety of technical approaches and process variations have been or are being pursued by different groups, and these are listed in Figure 5. The reactor types include moving beds; fluid solids systems with single or multiple stages; reactors with ash agglomerating, ash slagging, or dry ash removal features; and molten salt or molten iron baths. Some systems also use catalyst or dolomite addition. The methods for heat input include oxygen injection directly into the gasification bed, air combustion outside the gasification bed, electric heat, and the recirculation of hot steams of gas or solids. Gasifier pressures range from atmospheric to about 1500 psi, and gasifier temperatures range from about 1300 to 3000°F. Altogether, many combinations of gasifier types and operating conditions have been or are being pursued.

- o Commercially demonstrated technology is available --- Lurgi, Koppers-Totzek, Winkler
- o Groups in U.S. have announced plans for Lurgi SNG projects, but none have reached construction, the problems have been
  - Government approvals and regulations
  - Environmental clearances
  - cost and pricing of gas
  - Financing arrangements
- o New technology will not overcome these barriers

Figure 3. Gasification technology

- o Two main objectives -- Reduce cost; Process a wider range of coal types and particle sizes
- o Process improvement goals to achieve lower costs
  - Reduce heat input to gasifier (produce more methane directly in gasifier, reduce temperature)
  - Accomplish heat input without pure oxygen
  - Reduce equipment multiplicity
  - Reduce number, complexity, & size of process steps
  - Improve heat recovery & utilization
  - Improve operability & reliability of overall system

Figure 4. New or improved coal gasification processes for SNG

And now I would like to comment on how much cost reduction can we really expect from all of this work. Figure 6 shows a breakdown of investment by plant section for a typical Lurgi SNG plant. The information is about three years old from the open literature. As shown, the gasification section accounts for only about 20% of the total plant investment. Other process sections, including shift, methanation, and other process gas account for another 30%, making a total of 50% for the process sections. The utilities add up to 33%, including 11% for the oxygen plant alone.

What this means is that any cost reduction in the gasification section alone cannot have a major impact on the overall gas cost. For example, a one-third reduction in the gasification section investment would reduce the total investment by about 7%, one-third of 20, and this corresponds to a reduction in gas cost of only about 3-4%. Therefore, any improvements in the gasification section should be aimed at reducing costs in the other plant sections as well. This conclusion was the basis for my earlier description of process improvement goals. As you may recall, I highlighted a number of items that could have their main impact outside of the gasification section. One of these was to accomplish the heat input without pure oxygen which would eliminate the oxygen plant. Another item was to produce more methane directly in the gasifier which would reduce the size, or change the nature, of the downstream process sections.

From the work that we have done, we have drawn certain conclusions about the potential for cost reduction. These are outlined in Figure 7. For the new thermal processes that have been studied the most in recent years, we find it difficult to see how the first commercial plants can provide much more than about 10% reduction in SNG cost over existing technology. As a mature industry is developed, and additional plants are built for the individual new processes, an additional 10-15% cost reduction might be achieved for the really good new processes. This additional reduction would require that further improvements be developed from the commercial plant operating experience and from continuing R&D. Altogether these are certainly worthwhile cost reductions, but they should not greatly affect the overall economics of plants built earlier using existing technology, such as Lurgi. The earlier plants should be able to continue operating viably for normal project lives.

For catalytic gasification, we believe that the potential for cost reduction is greater. For example, for the SNG process that we are now developing with DOE for bituminous coal, and for a pioneer plant, we estimate a potential reduction in gas cost over existing technology of about two times our estimated reduction for the thermal processes. In this regard, we have not had the opportunity to evaluate other newer processes that involve

REACTORS

- o Moving Beds
- o Fluid Solids -- single or multiple stages
- o Ash Agglomerating/slagging/dry ash removal
- o Molten salt or molten iron baths
- o Catalyst addition
- o Dolomite addition

HEAT INPUT

- o Oxygen or air injection into gasifier
- o Air combustion outside gasifier
- o Electric Heat
- o Recirculation of hot gas/solids

PRESSURES -- ATMOSPHERIC TO 1500 psi

TEMPERATURES -- 1300 to 3000°F

Figure 5. Variety of technical approaches

	%	
COAL HANDLING	8	
GASIFICATION	20	PROCESS SECTIONS, 50%
SHIFT & METHANATION	9	
BY-PRODUCT RECOVERY	7	
GAS PURIFICATION	11	
SULFUR PLANT	3	
OXYGEN PLANT	11	UTILITIES, 33%
STEAM & POWER	17	
WATER	5	
SITE, BUILDINGS, ETC.	9	
	100	

Figure 6. Breakdown of investment for typical Lurgi SNG plant

very short reaction times achieved through the use of rocket technology. Therefore, I cannot comment on their potential for reducing SNG cost. I will come back to these processes later on.

So far, I have talked mainly about cost reduction but what is the cost of SNG produced from coal? Figure 8 show a summary. There is certainly a wide range of views and numbers that have been published. Depending on the bases and the accounting methods used, SNG costs ranging from $3 to $7/MBTU, in 1978 dollars, have been quoted. The upper half of the range $5-7, is, in our opinion, probably more realistic, particularly when feeding higher-cost, deep-mined coals.

In this regard, industry and government have had a track record of generally under-predicting synthetic fuels costs. Some of the factors contributing to this include the following: optimistic process predictions based on limited data; incomplete development of all process features; limited depth of engineering design; weak definition of support and offsite facilities; weak project definition; and, finally, the inexperience of many of the study groups in preparing cost estimates for very large and very complex projects. Altogether, it is very difficult to arrive at realistic cost estimates for a complex new technology.

In this regard, even the first commercial application of a new process can have substantial technical and cost uncertainties if the development program has not been carefully planned and conducted. This is outlined in Figure 9. One of the main challenges in planning the development program is, first, to determine whether a large pilot plant is needed, and then, if it is, to establish the proper design and size for this pilot plant. The main purpose of a large pilot plant is to provide the engineering scaleup data that cannot be obtained in smaller equipment and which are necessary before a commercial plant could be designed with normal technical risk. This development approach, if properly carried out, can eliminate the technical need for what is called a demonstration plant which is a plant intermediate in size between the large pilot plant and the commercial plant.

Establishing the proper design and size for the large pilot plant is easy to say but difficult to do. It is difficult because it requires the developer to prepare a projected commercial design first, and then to work backwards to determine the proper large pilot plant design. This is done by careful engineering analysis of each section of the projected commercial design to determine two things: (1) what scaleup data will be needed to prepare the definitive commercial design; and (2) what is the minimum size pilot plant and the proper design of this plant that can provide these data with a reasonable operating program.

- o For new thermal processes studied the most in recent years
  - First commercial plants, about 10% reduction
  - Subsequent commercial plants, an additional 10-15% reduction for some processes
- o Should not greatly affect economics of plants built earlier using existing technology
- o For catalytic gasification
  - Potential for cost reduction is greater

Figure 7. SNG cost reduction

- o Wide range of views and numbers have been published
  - $3 to $7/MBTU (1978 Dollars)
- o Upper half of range probably more realistic, particularly when feeding higher-cost, deep-mined coals
- o Industry and government have generally underpredicted synthetic fuels costs
  - Optimistic process predictions
  - Incomplete development of process features
  - Limited depth of engineering design
  - Weak definition of support & offsites facilities
  - Weak project definition
  - Inexperience in cost estimating of large and complex projects

Figure 8. Cost of SNG from coal

If this analysis is not made before the large pilot plant is designed, then the pilot plant can easily become an unfocused and drawn-out "trial-and-error" operation that will not provide the necessary scaleup data. In such a case, the subsequent commercial plant, or demonstration plant, if it is ever built, would itself become a very large pilot plant in many respects. Unfortunately, this can lead to excessive down times, fixup costs, operating failures, and other difficult situations in the commercial or demo plant. The overall result can be a very unsatisfactory and perhaps a disastrous project. This is why we believe so strongly in proper planning and conduct of the overall development program, including the large pilot phase.

And now I would like to comment on the National R&D effort on producing SNG from coal. Figure 10 shows a summary. There is a growing appreciation that the true cost of producing SNG from coal will be high. There is also uncertainty about the impact of the recently passed natural gas act on natural gas supply. And as Art mentioned earlier today, additional natural gas may become available from unconventional sources, such as tight formations and geopressured aquifers, and the Department of Energy is funding R&D work in these areas. Regarding new SNG processes, DOE has been considering which demonstration plant project to fund: the slagging Lurgi, COED/COGAS, both, or neither. DOE has also awarded a contract to procon to prepare designs for a conceptual Hygas process. This may or may not influence their considerations on the Lurgi and COED/COGAS plants. I'm sure that we will hear more about this later today.

Regarding DOE's large pilot plants, the synthane plant was recently shut down. The Hygas plant, as we understand it, is scheduled to operate through June, 1979 to provide backup for Procon's design work. Bigas will operate through September, 1979 and possibly beyond.

DOE is also funding research on newer gasification processes, sometimes called "third generation" processes. One of these is Exxon's Catalytic Coal Gasification, or CCG, process. In CCG, we use a potassium catalyst in a fluid bed gasifier. The catalyst allows us to operate at lower temperatures and to produce a high yield of methane directly in the gasification reactor. The methane product is then separated cryogenically from a recycle stream of CO and H_2 which is returned to the gasification reactor to help produce more methane and to provide heat input.

The other two processes, by Rockwell/Cities Service and Bell Aerospace, are based on Rocket Tehcnology. They both utilize high mass flux reactors in which finely powdered coal is rapidly fixed with high velocity, hot gas. The mixture is then quickly quenched to give very short reaction times. The Rockwell Process

- One of the main challenges is to...
  - Determine if a large pilot plant is needed
  - If so, establish proper design and size
- Main purpose of large pilot plant is to provide engineering scaleup data for commercial design
- Must do a commercial design first and then work backwards to proper large pilot plant design
  - Requires careful engineering analysis
- If this is not done properly
  - The large pilot plant could be ineffective
  - The subsequent commercial or demonstration plant could have major problems

Figure 9. Planning the development program

- Growing appreciation of true cost of SNG
- Uncertainty of impact of natural gas act on supply
- DOE if funding R&D on unconventional sources
- Demonstration plant competition
  - Slagging Lurgi and COED/COGAS
- DOE's large pilot plants
  - Synthane shut down
  - Hygas operation scheduled through June 1979
  - Bigas operation scheduled through September 1979
- DOE also funding newer processes
  - Exxon catalytic coal gasification (CCG)
  - Rockwell - Cities Service
  - Bell Aerospace

Figure 10. National R&D effort on SNG from coal

reacts the coal with hydrogen aiming at SNG. The Bell Aerospace Process reacts the coal with oxygen or air aiming at medium or low Btu gas.

Altogether, it would appear that DOE's overall commitment to gasification R&D has not decreased, although the National record of success for developing lower cost SNG processes has not been particularly outstanding.

And now the last area on which I would like to comment is environmental considerations. A summary appears in Figure 11. The environmental aspects of coal gasification plants could become a major issue, both technically and politically. Technically, the environmental requirements and water availability and consumption could play major roles in determining where SNG plants will be located and what the gas cost will be.

The technology is now available for cleaning up gas and water effluent streams, for controlling particulate emissions, and for minimizing water consumption. For reasonable requirements in these areas, the total cost in an SNG plant for gas, water and particulate cleanup would be roughly 15-20% of the total plant investment. This is clearly a major cost, but at this level it would not be an overwhelming cost. However, the cost would increase very rapidly if "clinical purity" were to be unnecessarily imposed on SNG plants, and the cost would indeed become overwhelming. This is a key area, and we hope that reason and good judgment, rather than emotion and unjustified imposition, will prevail.

- Environmental aspects could become a major issue
- Environmental and water considerations could play major roles in determining plant locations and costs
- Technology is now available for cleaning up gas and water effluents, controlling particulates, and minimizing water consumption
  - Cost is 15-20% of total plant investment
  - Cost would increase very rapidly if "clinical purity" were unnecessarily imposed
- Hopefully, reason and good judgment will prevail

Figure 11. Environmental considerations

RECEIVED May 21, 1979.

Major Technical Issues Facing Low and Medium Btu Gasification

E. L. CLARK

4615 North Park Avenue, Chevy Chase, MD 20015

The title of this paper requires some discussion to picture properly the status of coal gasification. An appreciable number of commercial coal gasification plants are operating in several countries throughout the world using several gasification processes. This status indicates that a reasonable number of technical issues have been solved and that a fundamental technical basis for coal gasification exists. Problems still facing the commercial use of coal gasification include the adaptation of existing processes to our environmental standards and to coals of United States origin. The problems of economics are also serious issues which are partly technical in nature. Process improvement and new process development are the technical issues we face in achieving economically competitive coal gasification.

Even though Low Btu gas (LBG) and Medium Btu gas (MBG) have become terms of common use, some specification of these gases is desirable. A brief specification is provided in Table 1 which is intended to cover LBG and MBG. The upper and lower values of Btu content per standard cubic foot should be considered as approximate rather than exact limitations. Similarly, the term "essentially free" is an attempt to avoid predicting what purity environmental standards might require in the future. The advantages and difficulties of gasifying coal at elevated pressure are not always appreciated. While combustion of LBG and MBG may take place at essentially atmospheric pressure, the generation of these gases at elevated pressure can provide more economical gasifier operation and more convenient transport to several users. Finally, if a clean gas could be furnished at elevated temperature, the thermal content of the gas would be available to the user.

While the specification in Table 1 covers both low and medium Btu gases, we are discussing two different materials. Low Btu gas (LBG) is produced by the reaction of air and steam with coal and has a heating value generally 150 to 170 Btu/Standard Cubic Foot (SCF). Medium Btu gas (MBF) is produced by the reaction of

0-8412-0516-7/79/47-110-183$05.00/0

oxygen and steam with coal and has a heating value of 290 to 350 Btu/SCF. The difference in heating value is critical. Conversion of an existing oil or natural gas-fired steam generation or process heating unit becomes quite costly when the alternative gaseous fuel heating value drops below 250 Btu/SCF.(1) This indicates that end-uses of LBG may be limited to new plants designed specifically for gaseous fuels of low heating value.

TABLE 1

LOW BTU GAS SPECIFICATION

- o Heating value above 120 Btu/SCF and below 500 Btu/SCF
- o Essentially free of sulfur, ammonia, particulates and hazardous impurities or byproducts
- o Preferably provided at elevated pressure
- o Preferably provided at elevated temperature

Another important difference is that the complexity and minimum economic size is considerably greater for an MBG plant than an LBG plant. Small LBG plants are on operation supplying gaseous fuel to small industrial plants.

Bearing these points in mind, we can consider the potential markets for LBG and MBG. Table 2 provides a listing of industrial fuel usage and power generation supplied by petroleum and natual gas. The fuel or energy amounts are in Quads and we can conveniently picture the size of a Quad by noting its equivalence to one trillion cubic feet of natural gas or 10^{15} Btu. The data for 1974 are approximately the same for the period of 1974 to 1977. The data for the year 2000 were taken from a projection made some time ago by the Electric Power Research Institute (EPRI) and appear a bit on the high side for projected increases for electric power and total energy for the year 2000. In any case, we can conclude that a sizable market potential exists for MBG as an alternative fuel for existing units. Similarly, the growth projections for the future indicate an adequate potential for LBG as a fuel for new facilities especially for electric power generation. We caution that the growth projection to 2000 given in this Table is quite tentative and several other projections indicate lower total energy demand by that year.

The chemical reactions taking place during the gasification of coal are well known. Some of these are listed in Table 3. In the reactions listed, coal is assumed to be essentially carbon. The oxygen is either pure oxygen as used in MBG production or oxygen contained in air for LBG generation. In the latter case, nitrogen will be present as a diluent. The first three reactions listed are truly gasification reactions in that they convert a solid (carbon) to a gas. It is apparent that the reaction of carbon with oxygen must supply all the heat energy required.

TABLE 2

POTENTIAL LOW BTU GAS MARKETS

	1974 QUADS	2000 QUADS[3]
Industrial	16.0[1]	
	(20.4)[2]	(30)[2]
Electric Power	7.0[1]	
	(20.0)[2]	(75)[2]
TOTAL	73[2]	150[2]

(1) Market supplied by petroleum and natural gas.

(2) Total demand.

(3) Estimates prepared by EPRI.

TABLE 3

COAL GASIFICATION KEY REACTIONS

- Gasification

  $C + H_2O \longrightarrow CO + H_2$; Endothermic

  $C + O_2 \longrightarrow CO_2$; Exothermic

  $C + 2H_2 \longrightarrow CH_4$; Exothermic

- Shift

  $CO + H_2O \longrightarrow CO_2 + H_2$; Exothermic

- Methanation

  $CO + 3H_2 \longrightarrow CH_4 + H_2O$; Exothermic

While some heat energy may be supplied by the hydrogenation of carbon to methane, the hydrogen required for this reaction must be supplied by the small amount of hydrogen in the coal or by the endothermic reaction of steam with carbon. Heat for this reaction must be supplied by combustion of carbon. The two gas phase reactions alter the composition of the gases produced. The shift reaction is slightly exothermic. The methanation reaction is strongly exothermic but requires the presence of hydrogen and either elevated pressure or an active catalyst.

As in most reactions between solids and gases, the method of obtaining contact between coal and reactant gases is a critical factor. Figure 1 shows three additional systems for solid and gas contacting. All three are used in coal gasification commercial units or are the basis for processes under commercial development. There are advantages and disadvantages to each system. The moving bed unit on the left uses fairly large sizes of coal with a minimum size of one-quarter inch. It provides countercurrent flow and good heat transfer. The fluidized bed shown in the center uses relatively small particles of coal which result in a more rapid reaction rate. Both moving bed and fluidized bed units have difficulties in handling coals which agglomerate. Both require precautions in preventing softening or melting of ash which may cause formation of clinkers and may disrupt solid flow. The entrained flow unit uses fine particles of coal; operates at higher temperatures to obtain rapid reaction rates; and removes ash in the molten state or as slag. This unit can handle any type of coal but attention must be paid to the ash components and the melting point and melt viscosity to obtain reliable operation.

The fixed or moving bed has had more usage than any other system. Many small units were operated here and abroad. Used with air and non-agglomerating coal and operated at essentially atmospheric pressure, such units were inexpensive, simple to operate and widely used. The Lurgi unit is the only one which has been designed for operation at elevated pressure. It can be used with air or oxygen with the latter more widely used. These units use modest amounts of oxygen (160 to 170 cu. ft. oxygen/1000 cu.ft MBG) but, in order to protect the grate which discharges the ash, use quite large quantities of steam (approx. 75 lbs/ 1000 cu.ft. MBG). In all fixed bed units, the hot gases flow upward heating and devolatalizing the coal which enters at the top. These volatiles condense and result in the production of oils, tars and various organic contaminants. The latter are found in the discarded water condensate and necessitate an expensive water cleanup system prior to disposal. A major problem is the need to find a reliable use for the fine coal which the fixed bed cannot handle. A very large pipeline gas plant which plans the production of MBG (for conversion to methane) from North Dakota lignite using Lurgi gasifiers has arranged to sell all the fine coal to a

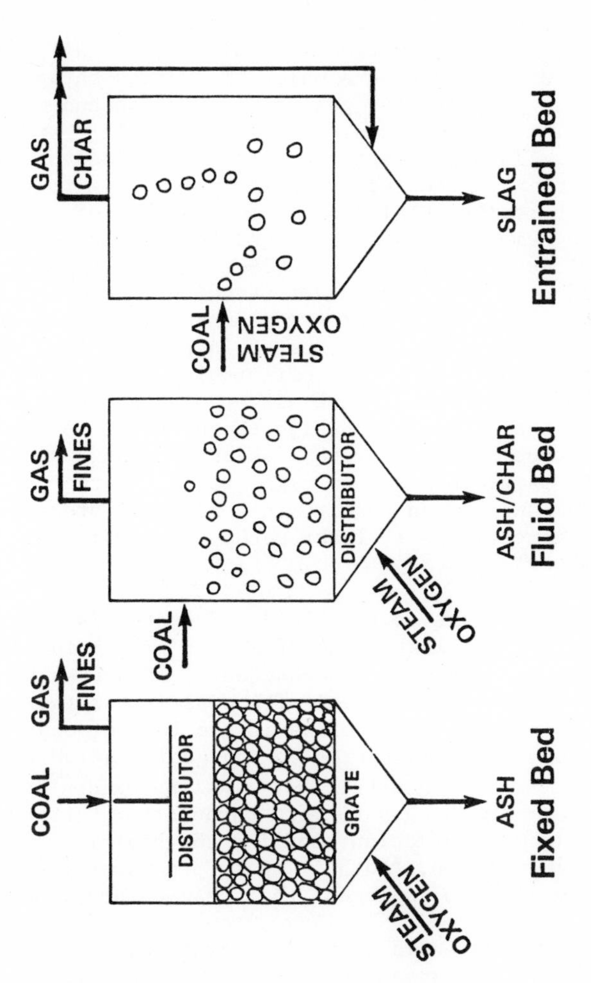

Figure 1. Major gasification systems

nearby lignite-based power plant concurrently under construction. Whether this pattern can be continued in all cases where fixed bed gasification will be used is doubtful.

The Fluid Bed System shown in Figure 2 is the basis for several developmental processes. The use of fine particles permits total utilization of coal-mine output. One commercial process, the Winkler process, is used in Asia and Africa to provide MBG for fuel and chemical synthesis. This commercial process is operated at atmospheric pressure which is a disadvantage due to compression costs required for gas transportation and most chemical end uses. The developmental processes are all operated at elevated pressure in an attempt to remedy this disadvantage. The fluidized bed, being a completely mixed system, limits the carbon conversion which can be obtained. As much as 15% of the coal is not reacted and some use must be made of the high-ash-content char. The use of agglomerating coal is precluded due to the loss of fluidization if coal particles start sticking together. The use of fine particles does permit pre-treatment of agglomerating coals prior to feeding to the gasifier, but this process also entails losses in carbon conversion. Another problem area is the lower portion of the fluidized bed where air or oxygen enters and first reacts with the coal. Localized high temperatures in areas where adequate turbulence of flow may be lacking can cause sintering together of ash particles to form clinkers and disrupt operation. Reasonable steam and oxygen requirements may make processes based on this system competitive if lower carbon conversion can be tolerated.

An important variant of the Fluid Bed system is under development. This variant eliminates use of air or oxygen in the actual gasifier. Steam and coal are the reactants. Since we know from Table 3 that the reaction of steam with coal is endothermic, a heat source must be provided. Hot solids in the form of char are heated in a combustor and are transferred to the gasification reactor as one these processes. In another, hot alkaline oxides react with the carbon dioxide in the gas to form carbonates. The exothermic reaction of carbonate formation supplies the heat requirements of the steam-carbon reaction. Both of these processes depend on a reactive coal or char to implement the steam-carbon reaction.

The Entrained system is a high temperature, high reaction rate process in which coal, oxygen (or air) and steam combine rapidly to produce LBG or MBG. The commercial processes aim primarily at the use of oxygen. Several developmental processes use oxygen or air. The most widely used commercial process (Koppers-Totzek) is operated at atmospheric pressure. The Texaco partial oxidation process used with oil and gas is under development for use with coal. Shell and Koppers are developing a pressurized version of the current Koppers-Totzek process. The advantages of the entrained

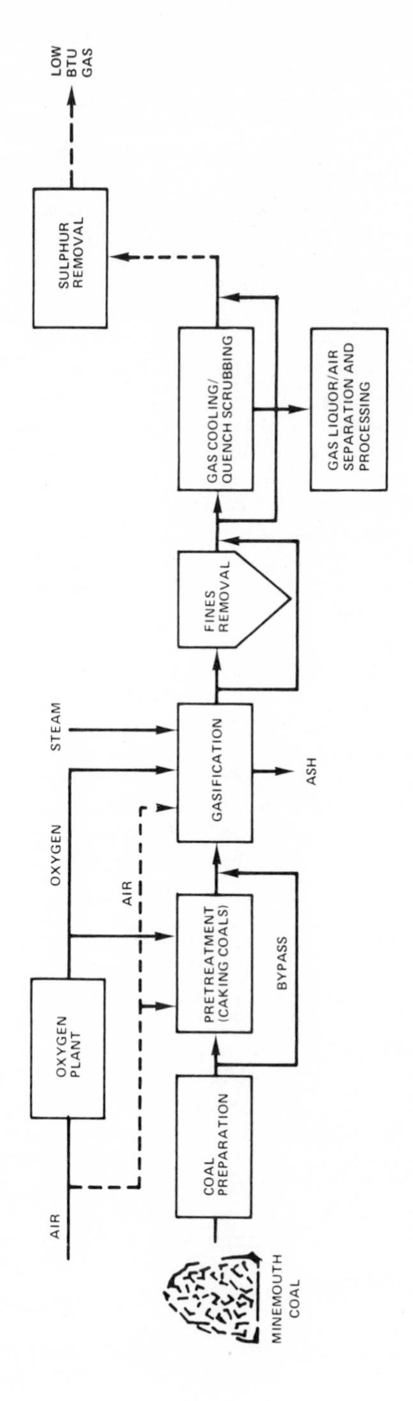

Figure 2. Anatomy of gasification processes, low Btu version

systems are: complete conversion of coal; ability to use almost any coal, agglomerating or not, and of almost any rank; and an apparent lack of adverse environmental impact since no oils, tars or contaminants are formed. Problem areas involve: control of particulate emissions; handling of molten slag and need for suitable refractories; and, due to high exit temperatures from the gasifier, a need to develop suitable heat recovery systems. The entrained systems are generally high oxygen consumers - almost double the requirement for the fixed bed units. Energy for oxygen production could be recovered from hot gasifier effluent gases if suitable waste heat boilers and superheaters can be developed.

To this point, we trust that a clear picture of the technical status of coal gasification is emerging. We have a reasonable grasp of the chemistry. Three systems for handling the mechanics of coal reaction with steam and air/oxygen have been developed to the point where commercial operation is practiced. However, this achievement is not complete enough for widespread commercial use in the United States. The technology must be made to conform with environmental standards, economics and the end-use patterns of potential customers. No meaningful demonstration of coal gasification technology has been provided to establish this requirement and to prove operability and reliable on-stream performance. Until this is achieved, economic estimates degenerate into inconclusive paper studies and potential customers cannot accept the risks involved. Until one or more state-of-the-art systems are operated on a commercial scale, the most attractive advanced systems cannot be moved further toward commercialization except through massive subsidies by the Department of Energy (DOE). These do not seem to be available in today's stringent budgets which are aimed at an elimination of deficits and a reduction in inflationary impact.

Some steps forward are being made in establishing real costs, collecting environmental information and demonstrating reliable operability. Several small fixed bed gasifiers sponsored by DOE and industry are under construction. These will produce LBG for industrial use. All of these are air blown units and are state-of-the-art gasifiers. A sizable environmental evaluation and control program is being implemented. Within the next 12 to 18 months we should have operating and reliable economic data on these systems. While the impact on national energy usage of these relatively small units be negligible, the data provided will establish a lower technial risk level for larger fixed bed units. Both effluent control and control of in-plant toxic substance levels will be reduced to industrial practice and should make future plants easier to build.

The economic position of these small units is not very advantageous. In certain end-uses, hot unpurified gas may be utilized directly to the kiln burners. Particulate control is necessary at the kiln outlet in any case and the small amount of coal ash carried over from the gasifier does not seem to affect brick product quality. A similar system is being tested at the Bureau of Mines Twin City Station in St. Paul, Minnesota. Here LBG is produced and used to sinter taconite pebbles. The process is called endurating or hardening. Again, the hot gas from the gasifier is fed directly to the shaft furnace or kiln to fire the "green" taconite pebbles and harden them. In this case, the small amount of sulfur in the gas being produced from lignite is absorbed by the iron oxide pebbles with no serious effect on quality. Particulate control is maintained at the outlet of the processing kiln or furnace.

In cases such as the two described, we can visualize a competitive position for LBG. Cost estimates for these "hot, dirty gas" generation systems show a fuel cost of under $3.00/million Btu in 1976 dollars. However, when a purification system for both particulate and sulfur removal is added to these small-size production units, the cost increases drastically. The average output of these small, air-blown gasifiers operating at atmospheric pressure is less than 10 tons of coal/hour. Single train purification systems can handle the gas production from as much as 5,000 tons of coal/day. It is obvious that such large systems are much less costly per unit of production than a small unit handling the gas produced from 200 to 250 tons of coal/day. Another disadvantage in purifying the gas from these saml1 units is their operation at atmospheric pressure. The smaller volume of gas at elevated pressure further reduces the capital cost of purification systems. As a result, one finds that purified LBG in small units may double the price to over $5.00/million Btu. Thus, LBG in small units is only competitive in rather special cases.

For larger units using 10,000 tons of coal per day, costs of purified gas suitable for combustion under even the most stringent environmental criterion are becoming competitive. Costs for producing gas by state-of-the-art fixed bed systems operated at 300 psig using coal costing $1.00/million Btu have been estimated by EPRI at $3.00 to $3.50 for LBG and $3.50 to $4.50 for MBG with all values in dollars per million Btu in mid-1975 dollars. (2) The variation in cost is primarily a function of the operating factor which might be assumed. This was varied from 70% to 90%. It should be realized that these are very large plants producing slightly more than 130×10^9 (billion) Btu/day. This quantity of energy could generate in excess of 640 MW of electricity (assuming a heat rate of 9,000 Btu/KW). Finding an industrial plant large enough is not easily achieved even in today's policy of very large industrial production units. A survey made for the President's Energy Policy and Planning Office (3) in 1977 shows fewer than 20 plants large enough in natural gas consumption to use the output of an MBG or LBG plant which could process 10,000 tons of coal/day.

We are forced to conclude that LBG or MBG generated in fixed bed units could approach commercially competitive levels in large plants. For LBG, which cannot be conveniently transported, only very few industrial plants could justify on-site generation at an economic scale. An LBG plant processing 10,000 tons of coal/day could provide energy translatable into 650 to 800 MW of electricity. Remembering that conversion of existing facilities to LBG is expensive and difficult, we find a relatively minor role for industrial use except in new, large plants using in excess of 15×10^{12} (trillion) Btu/year or in large power generating plants of over 500 MW in size. One must expect that electric power generation offers the greatest potential for LBG and some future thrust at commercialization might be sponsored or activated by the public utility sector.

For MBG, industrial use in large, existing plants has a meaningful potential. Of the large plants which might support an economically-sized MBG plant, over half are petroleum or petrochemical facilities. Implementation of MBG to supply fuel and gaseous feedstocks to such plants would almost directly reduce petroleum consumption in such facilities. This reduction would be appreciable since it is estimated that 6% to 10% of the crude petroleum fed to a refinery might be utilized to provide energy for the refining process. A study made for DOE has indicated several areas where suitable concentrations of industrial plants could be served by a single MBG gas producing facility. (4). Due to the need for transporting the gas, LBG could not be used. Examples of such areas are: Houston, with a need for 149×10^{12} Btu/year by 1985; Chicago, with 69×10^{12}; Pittsburgh, with 25×10^{12}; St. Louis, with 20×10^{12}, and Philadelphia, with 37×10^{12}. These five areas represent over half the total United States potential requirement for MBG fuel. Individual facilities in each of these areas could supply MBG to many industrial plants.

While the evaluations of cost and plant size discussed in the preceding paragraphs have been devoted to fixed bed systems, the conclusions are valid for all coal gasification techniques. Estimates of fluid bed gasifiers have also been prepared. (2) Unfortunately, insufficient data are available to substantiate the operability and actual productivity which must form the basis for any cost estimate. Using these tentative costs, we find that costs for LBG or MBG might be below $3.00/million Btu. For entrained flow systems, still under development, costs in the below $3.00/million Btu range are estimated. However, actual implementation of these advanced systems or even state-of-the-art systems is still a matter of willingness to take appreciable technical and economic risks. The costs of MBG and LBG should become competitive with energy prices in the 1983-85 time span as petroleum prices increase and natural gas prices are decontrolled. Further development of improved gasification processes should also have some effect.

The effects of new process development are, however, limited to actual gasification and these may be small. The total facility for producing LBG or MBG is somewhat more than just gasification. Gas purification, waste disposal and general utility requirements are almost all standard systems which will only be partially reduced in cost by improved coal gasifier technology. Figure 2 diagrams the units required to produce LBG or MBG. In the case of LBG, air is utilized bypassing the oxygen plant which is required for MBG production. Gas cooling, fines removal and sulfur removal are similar for producing both gases. Similarly, coal preparation and pretreatment are performed in similar systems for both gases.

The additional technical barriers which must be overcome have been stated. Primarily, the need is greatest for actual operation and demonstration of gasification on an industrial scale. Small gasifiers are being so demonstrated through the assistance of the Department of Energy (DOE). Additional efforts are underway in the DOE program. These include an MBG demonstration plant in which one of two processes will be tested: either production of ammonia synthesis gas; or production and distribution of a fuel gas to several industrial and power generation customers. The generation of MBG may be also demonstrated in the pipeline gas demonstration program. While the MBG produced under this program will be converted to synthetic natural gas, the generation of MBG demonstrated in a pipeline gas plant could be also applied to producing industrial fuel or synthesis gas. One similar LBG demonstration plant in the DOE program will use LBG for enduration of taconite pellets. While implementation of these demonstration plant projects will depend on the magnitude of the DOE budget, a very large share of the gasification budget is being committed to this effort.

Fortunately, efforts in addition to those of DOE are being implemented. The Tennessee Valley Authority is sponsoring the construction of an entrained flow gasifier to operate at elevated pressure and to provide synthesis gas to their small Ammonia Plant at Muscle Shoals, Alabama. Ironically, this ammonia plant was originally built using coke-fed water gas sets for synthesis gas production. It was converted to use natural gas steam reforming when cheap natural gas became available. The use of coal will provide valuable data on MBG production and purification. The Carter Oil Company has reported its studies on using

Texas lignite to generate MBG which would be piped to the Houston area for fuel and feedstock use. These studies have included the testing of Texas lignite in commercial gasification units located abroad.

A very large gasification project for converting coal to MBG and pipeline gas is under consideration by the Federal Energy Regulatory Commission with a decision expected by June 1979. Implementation of this project would provide a major demonstration of MBG production and purification. The design effort for this project has included large-scale tests of North Dakota lignite in commercial coal gasification units. It is anticipated that projects of this magnitude, when successfully operated in the 1983-85 time period, will provide sufficient data so that normal industrial decisions on use of MBG or LBG can be made. The technical risk should be minimized to permit normal financing.

Several important development efforts could improve the economic status of low Btu gas production. Tests performed at Westfield, Scotland, jointly by the Energy Research and Development Administration (ERDA) and the American Gas Association (A.G.A.) (5) have demonstrated that fixed bed gasifiers can be used successfully with weakly caking coals (up to a free swelling index of 2.5 to 3.0) if suitable stirrers and distributors are utilized. While small-scale tests at the Morgantown Energy Technology Center have demonstrated on a pilot plant scale that even highly caking coals can be handled, these tests partially confirmed the Morgantown results on commercial-size gasifiers. More recent results with fixed bed units at Westfield have demonstrated the operation of a slagging bottom instead of a grate. This could reduce costs appreciably for fixed bed gasification by reducing steam requirements used for grate cooling by over 90%. Further, longer-term demonstration of the operating feasibility of this improved gasifier appears desirable. A major problem in fluidized bed gasification is the low carbon conversion. Ash agglomeration could improve carbon conversion and use the fines effectively. Test work on this system is in progress on a Process Development Unit scale. Finally, the use of entrained systems at elevated pressure should improve their applicability to a greater variety of end uses.

The implementation of coal gasification will occur as more data are available to eliminate technical risk. Currently we can visualize a competitive cost or $3.00 to $3.50/million Btu for LBG and MBG in large units utilizing 5,000 to 10,000 tons of coal per day. These units could provide a guaranteed supply of gas to industry without being diverted to use for priority consumer needs. MBG particularly could become a distributed gas for industrial use. Several areas where suitable industrial plants are concentrated have been listed in a study sponsored by

DOE (4). The implementation of one or more demonstrations of coal gasification by 1983 to 1985 should provide a solid basis for commercial use. This represents an unusual opportunity for the gas industry to extend its operating base and to ensure future supplies of clean fuel for consumers and industry.

A similar opportunity exists for the public utility industry in the potential of LBG and MBG. The reduced environmental impact of a coal gasification plant which produces a perfectly clean fuel equivalent to natural gas, compared to direct combustion of coal may allow increased use of coal in areas where increased pollutant emission is barred. As these PSD areas increase in number, the advantages of coal gasification become more apparent. The potential of more efficient combined cycle generation systems which can be used with coal-derived gases is an added factor for implementing coal gasification.

"LITERATURE CITED"

(1) Low Btu Gas Study, Electric Power Research Institute, Report No. EPRI 265-2, January, 1976.

(2) Economics of Current and Advanced Gasification Processes for Fuel Gas Production; Electric Power Research Institute Report No. EPRI-AF244, July, 1976.

(3) Market Potential for Low and Medium Btu Gas, Energy and Environmental Analysis, Inc., November 4, 1977.

(4) Market Opportunities for Low and Intermediate Btu Gas from Coal in Selected Areas of Industrial Concentration, SRI International, Report No. HCP/T2441-02, June, 1978.

(5) Trials of American Coals in a Lurgi Gasifier at Westfield, Scotland, Report No. E-105, December, 1972-November, 1974.

RECEIVED July 11, 1979.

12

Energy Commercialization Prospects

RICHARD A. PASSMAN

U.S. Department of Energy, 1200 Pennsylvania Ave., N.W., Room #3442, Washington, D.C. 20461

I am pleased to be here today to give some insight into the Department of Energy efforts directed to the commercialization of coal conversion processes for gaseous and liquid fuels. There can be no doubt as to the importance the DOE assigns to its coal activities when the budget for 1979 and that proposed for 1980 is about $700 M. To give emphasis to the administration's continued interest in coal use and development, I would like to quote some people of significance. At the signing ceremony of the National Energy Act last November, President Carter stated "we must shift toward more abundant supplies of energy than those that we are presently using at such a great rate: to coal with which our nation is blessed..." and on January 22 of this year, Dr. Schlesinger stated in a letter to Carl Bagge, "...let me re-emphasize that the administration has never veered, and is not now veering from its commitment to coal".

There appears to be recognition by all responsible people that oil and gas, as finite resources with ever-increasing worldwide use, will soon be depleted. Only the date of this occurrence and the timing of the developing shortages evoke varying shades of opinion. From a commercialization view, we are concerned with the economics as shortages approach, and have assumed a steady depletion of resources until the demand overcomes the available supply. This gradual approach may be unrealistic as we note that sudden events, such as occurred in Iran, can disrupt this "schoolbook" case, and can change the economics by a sudden denial of the resource, by the world bidding up the price of a resource in short supply, possibly to unacceptable levels. Other effects stemming from abrupt reduction of supply include excessive negative balance of payments and pressure on certain foreign policy decisions and national security matters.

For all of these reasons, there is increased attention on coal conversion as a means of using our abundance of coal to supplement and replace other energy sources.

In DOE, we've come a long way since the beginning in October 1977 when the state of readiness for commercialization of any technology was undetermined, and where there was no system ready to evaluate or people prepared to make such judgments. Early in 1978, Undersecretary Dale Myers instituted a procedure to evaluate the technologies for commercialization potential, formed a commercialization committee to make the judgments, and later established the organizational position of resource manager to commercialize an assigned technology.

During that time, the department analyzed approximately one hundred technologies to determine which were ready for immediate commercial considerations. Commercialization task forces were formed for each technology of interest and briefings of their findings were made to the Commercialization Committee, where principal issues were explored in detail. The committee then directed the future task force efforts towards areas of their explicit interest. The preliminary conceptual phase was followed by an evaluation phase covering the market, the competition, the technical state of readiness, the important economics of capital cost and operating cost, environmental issues and institutional problems. A third phase concentrated on defining the barriers to commercialization and the federal actions that could potentially overcome the barriers.

The Commercialization Committee selected about 15 processes to be commercialized, and assigned a resource manager to head each one. Operating much like product managers in industry, they are the focus of departmental efforts to commercialize, to overcome the barriers, and to institute appropriate federal actions. The resource manager's objective is to establish a commercial capability, an experience base that provides capital costs and operating costs, maintenance, training, overall system performance ranging from transporting coal to waste disposal, and environmental suitability. The establishing of the experience base of operating commercial plants is to be performed and managed by the commercial community with as little government participation as possible. To be credible as being commercial, it should have minimum or no government participation. The major benefit to be gained by the federal government is the use of the resulting data to make it widely available to the potential using community, thereby accelerating its use.

We recognize that the U.S. Government can declare a technology as being commercial, but that no one may use it. Only the user and the suppliers can make it commercial by their actions: this cannot be done by federal decree. It falls upon the industry and the public to determine its commercial utility. Where necessary, however, the Federal Government intends to assist industry in making this possible. In some cases, this could be done by providing an operating plant example, by assisting with first-of-a-kind plant costs, with regulation determination, or with off-budget financial incentives such as investment tax credits or loan guarantees.

The job for the resource manager is formidable; it is not certain that even the best processes will become commercial in a competitive environment. Many external factors can cause the commercial environment to change with great effect. A few examples are: the availability of cheap Mexican and Canadian gas, an Iranian shutdown of oil production, a larger OPEC price increase than anticipated, very high interest rates.

Recognizing that our federal perceptions of what is commercial are limited, since we do not supply systems or compete in the marketplace, input is needed from the supplying industry, the users, and the participating financial institutions. The resource managers will need to develop sufficient data on plant costs, operation, etc., to make credible judgments. As the departmental focus on the changing environmental requirements and regulations that need to be sorted out, they will in some instances help to change them. The resource manager will be responsible for making commercial his assigned technology area.

One other factor of importance is the budget, and it is tight in a year where inflation-fighting has top priority. And I believe the budget will remain austere for at least another year. The commercialization objectives, therefore, must be accomplished with limited dollars. The government support will be only that necessary to help initiate a commercial capability that can provide the data and serve as a guide to others in their specific applications.

I would like now to address the commercialization of coal gasification and liquefaction processes, including the Department's planned activities in these programs.

HIGH-BTU GAS

First, High Btu Gasification, where the technology to be first commercialized is that of Lurgi technology, and for which the need will arise when the current gas surplus ends and when the equivalent cost of natural fuels rises to that anticipated for High-Btu Gas. Because these factors are expected to happen, we believe it important that the coal gasification capability be generated at an early date so that significant quantities of pipeline quality gas can be supplied by 2000 using domestic coal resources.

In the area of High-Btu Gas the pioneering efforts by the

Great Plains Gasification Association is to be commended. As you are probably aware, the department has intervened before the Federal Energy Regulatory Commission (FERC) in support of the Great Plains tariff requests. We believe that High-Btu Gas has broad applications and that through the network of transcontinental pipelines all sectors can be benefactors of supplies from this technology. The same product will be supplied to the same market--hence, no market analysis or changes in user equipment is needed. It is estimated that by the year 2000 the market for supplemental pipeline gas will be from 10 to 14 quads. High-Btu Gas has been estimated to supply a significant portion of the market at a levelized cost of about $4.00/Million Btu in 1978 dollars. If this be so, it would then be in the competitive range with other supplemental gas supplies, such as SNG from imported Naphtha, Alaskan Natural Gas and imported LNG.

Many of the uncertainties shrouding the development of this technology could be eliminated if we could provide actual investment costs, operating economics, environmental information, etc. However, without federal assistance--particularly some type of financial incentive--it appears unlikely that any commercial High-Btu Gas plants will be built. The federal role could be to assist the private sector in capital formation spreading the financial risk appropriately among project beneficiaries--be it industry, gas users or the public.

To achieve this goal, the department is currently pursuing a two-pronged approach to commercialization of High-Btu Gas. The initial effort is to support before the FERC a tariff mechanism that would enable a consortium, such as Great Plains Gasification Assoc. to finance a High-Btu Gas commercial plant. We are supporting rolled-in pricing for the coal gas, full recovery of debt plus interest, and partial recovery of equity capital: A second effort under consideration is to use a federal loan guarantee coupled with DOE support for an appropriate tariff before FERC.

The above program deals with the commercialization of First-Generation Lurgi technology. However, several privately funded projects and the Department of Energy in cooperation with numerous industrial groups, will be conducting extensive programs to develop improved coal gasification processes considered here as second generation. These improved processes, and they were discussed earlier, are intended to reduce the Synthetic Gas costs and to extend the application to eastern caking coals. But these second-Generation processes are not expected until the early 1990's as commercial.

Currently, we are reevaluating industrial interest, and hence the readiness of First-Generation High Btu Gasification, to recommend an appropriate timing of government actions to effectively stimulate industry. The timing of second-generation technology will become more important if delays occur in first-generation installations. Therefore, we are re-evaluating these processes as well, and generating the commercialization plans for each of them.

MEDIUM-BTU GAS

Now I would like to discuss Medium-Btu Gas. Many of the things applicable here are similar for Low-Btu Gas. The principal difference other than the Btu content between those two is the capital cost of the Medium-Btu Gas plant which is about an order of magnitude greater because of the size of plant needed for economic operation when using an expensive oxygen plant.

Medium-Btu Gas from coal (200-600 Btu/SCF) is a commercially available technology option for producing environmentally acceptable clean gas for both the industrial and utility markets. A total of 24 quads of energy utilizing natural gas and fuel oil are estimated to be consumed in these two markets by 1985. The primary markets for Medium-Btu Gas are as fuel gas for large industrial users such as the steel, refinery or chemical industries, as chemical feedstock; as a source of hydrogen for coal liquefaction; as fuel gas for utility application in combined cycle systems; or as a gas distributed regionally to a group of energy-intensive users through a closed-loop transmission system. Medium-Btu Gas is capable of being transmitted economically over an area of about a 50-100 mile radius depending, of course, on the size of the plant and the cost of distribution.

The processes and equipment currently available for producing Medium-Btu Gas include the Lurgi, Koppers-Totzek, Winkler, and possbily the Texaco process. Only Koppers-Totzek is able to process the eastern caking coals, but all can work on western coal. In spite of the commercially available technology, there are no Medium-Btu plants in this country. In contrast, there are as many as a hundred operating plants overseas.

In considering the option of Medium-Btu Coal Gasification, industry and utilities face major uncertainties and unknowns in properly assessing the technology and its utilization. Siting, distribution, coal supply, costs (capital and operating), reliability of operation, environmental compliance, retrofit problems, and acceptability of coal gas are some of the considerations which must be addressed. Initial commercial applications traditionally involve high business and investment risks. Planning a large multi-user Medium-Btu plant, one has to take into consideration the fact that problems, and needs vary by the industry application, geographical location, coal availability, and local regulations.

Another important consideration is that Medium-Btu plants require oxygen. If an oxygen plant must be built with the gasification unit, the economics of scale dictate that plants of larger than 30 billion Btu/day output would be necessary. A plant of 50 billion Btu/day is estimated to cost about $200 million. The cost of the clean gas produced at the gate has been estimated to be $3.75 - $4.50/MM Btu depending on cost of coal, capital and other factors.

The first step in our commercialization strategy for Medium-Btu Gas is to establish an experience base to provide industry the necessary information and confidence it needs to utilize this technology. We are planning to initiate actions which will support the design, construction and operation of several commercial Medium-Btu Gasification Applications by 1985. This assistance is to be limited to planning support such as establishing market requirements, feasibility studies, siting analysis, environmental assessments, and cost and financial analysis. This is to be offered to a number of potential users that have an interest in proceeding all the way to build a plant. We expect the notice of program interest to appear in Commerce Business Daily by the end of the month. There may be a subsequent program to provide assistance selectively to several promising applications from preliminary design through initial startup operation. This is not currently in the FY 1980 budget, but it is in our plan.

We are seeking to establish an experience base in categories that would include:

--- Large multi-user application
--- Industrial fuel application for chemical, steel and refining
--- Chemical feedstock application
--- Eastern and western coals and lignite
--- Several geographic areas involving attainment and non-attainment areas.

Another element of our commercialization strategy is determining means to stimulate and motivate industry through appropriate financial incentives, regulations, tax incentives, and federal policy. One last element is that of promoting industrial planning guides and conducting workshops so that appropriate industry members can evaluate their own situation for use of Medium-Btu Gas with the information base we plan to generate.

LOW-BTU GAS

In the area of Low-Btu Coal Gasification, the technology to produce environmentally acceptable (150-200 Btu/SCF) gas from coal is available. In fact, one can select from at least eight commercially available gasifiers to produce Low-Btu Gas. Gas clean-up systems are also commercially available but limited to a few systems. All of these systems are currently offered on a turnkey basis.

In the U.S. there are only two commercial users of Low-Btu Gasifiers operating today. In both cases, no sulfur removal and only limited gas clean-up is involved. At one time, (1920's) in the U.S. there were over 10,000 similar small gasifiers in use. But they were dismantled and put out of service as a result of cheaper, cleaner natural gas being available on a widespread basis.

Today, Low-Btu Gas (LBG) is expected to be preferred in small demand applications for single users located outside of downtown metropolitan areas. Specific industries in which LBG is expected to be most competitive include primary metals, iron ore beneficiation, metal finishing, lime brick refractory, and food industries. Another potential market for Low-Btu Gas is in combined cycle power generation. Cleaned Low-Btu Gas may be particularly advantageous when a plant has many separate combustors which, because of the anticipated new environmental standards, would require either multiple scrubbers or a flue gas collection system. Cleaned Low-Btu Gas is also one of the few options available to a user planning a plant expansion in a non-attainment area.

As in Medium-Btu Gas, we are planning to support planning assessments and feasibility studies this year to assist the development of several plants covering a range of applications. This will establish a commercial experience base and as in Medium-Btu, will provide industry the necessary information and confidence it needs to utilize this technology.

We will also perform an assessment of restraints to commercialization and evaluate appropriate incentives, and we plan to promote industrial interest through fact sheets, industry planning guides and by conducting workshops.

COAL LIQUEFACTION

In the field of coal liquefaction, many processes exist to convert coal to liquid and gaseous products. These processes can be categorized as direct and indirect liquefaction.

The direct liquefaction technologies, which include Solvent Refined Coal, Exxon Donor Solvent and H-Coal processes have never been operated at a commercial scale. As discussed yesterday, these processes are not at advanced stages of development. The products from direct liquefaction processes are basically boiler fuels or synthetic crudes that could potentially be upgraded to refined products.

The indirect liquefaction processes include Fischer-Tropsch and coal to methanol. Both processes have operated on a commercial scale. For the past 25 years, a Fischer-Tropsch facility has operated in South Africa. Presently the South Africans are constructing an advanced and larger facility. Coal-to-methanol plants existed in the United States, but were replaced by natural gas-to-methanol facilities because it was more economical to do so.

The earliest demands of the public for synthetic liquids will likely, in my mind, arise from gasoline shortages, causing lines at the gas pumps and restrictions on the use of the car. Transportation market (16.6 quads in 1976). Large market demands for liquid fuels also exist in the industrial and utility boiler and process fuel markets (5.4 quads in 1976). The indirect liquefaction processes produce products that are aimed directly at these markets. Methanol can be used neat as a transportation fuel in automobiles with modified engines as in racing cars. It can be blended with gasoline, requiring minor modifications to automobile engines, and thus act as a gasoline extender. Methanol can also be converted to high octane unleaded gasoline via a process being developed by Mobil Oil. Methanol is presently used as a petrochemical feedstock, and because of its clean burning characteristics has great potential as a fuel for power turbines, combined cycle, fuel cell, and boilers.

The Fischer-Tropsch technology produces a wide variety of products which can be narrowed to gasoline, diesel fuel, boiler fuel, distillate oil, and synthetic natural gas.

Because of the advanced stage of development of indirect liquefaction resource applications in DOE are aggressively pursuing the commercialization of the indirect processes.

Why is it, if indirect liquefaction processes are technically proven, the demand exists and is getting stronger for petroleum substitutes and there is so much coal available to us that people aren't standing in line to build coal liquefaction facilities in the United States today? The answer is fairly simple. There are so many uncertainties associated with commercialization -- not only technological, but also institutional, legal and regulatory-- that the large capital investments required seem too risky to make. Coal liquefaction facilities are capital-intensive with cost in excess of $1 billion.

The size of this investment, as well as the technological intensity, limits the number of companies capable of designing and efficiently operating coal liquefaction plants, and, as of now, neither the economics nor the long-term market potential is known.

The Department of Energy's Commercialization program is designed to identify the barriers, quantify them and provide the mechanisms necessary to hurdle them. The department feels that although estimates can be made of all the important commercial factors uncertainties will always exist until one or more plants are operated under U.S. market conditions at a commercial scale. It should be pointed out that although the South African Fischer-Tropsch facility is operating at a commercial scale, it is operating under very different market and regulatory conditions than exist in the U.S.

It is also produced with a different labor force, different automation and construction philosophy and with a different product mix. Likewise, coal to methanol plants that were built in the U.S. in the past did not have to contend with the environmental and institutional constraints that exist in the U.S. today.

In an effort to take the uncertainties out of the coal liquefaction industry and provide the confidence necessary before an industry will develop, the Department of Energy plans to support commercial ventures. Initially, support is anticipated for feasibility studies to identify, on a site specific basis, the economics, environmental requirements markets and feasibility for constructing and operating coal liquefaction facilities. It is intended that from these studies that detailed engineering designs, construction and operation of commercial-scale facilities will follow. We expect that the latter stages of these first commercial plants will not require direct federal involvement. However, off-budget incentives like accelerated depreciation, increased investment tax credits and loan guarantees are anticipated.

We intend to determine the commercialization advantages of the direct processes in comparison with the indirect processes discussed. The different markets, the relative economics, the state of relative development will all play a part in the recommendations planned to encourage the commercialization of the coal liquids technologies.

A commercialization program without industrial support and information is unthinkable. We solicit your thoughts, comments and recommendations on each of these activities. We welcome the opportunity to discuss your views on how best to provide the nation with a commercial capability on which industry can expand with confidence.

You have hands-on experience with the impediments associated with such large ventures. You also have the expertise in determining how these impediments can be overcome. Your assistance in defining how federal government actions can help overcome these barriers would be helpful and welcomed. Your support is needed to ensure that our program will be successful.

RECEIVED August 1, 1979.

13

Barriers to Commercialization

RICHARD F. HILL

Engineering Societies Commission on Energy, 444 North Capitol St., Washington, D.C. 20001

I would like to commence my discussion by giving you a very brief introduction to ESCOE. Some of what I am going to be saying has been developed by some of the people at ESCOE and I think it is important that you understand our perspective.

ESCOE is the Engineering Societies Commission on Energy which is a non-profit corporation that was established about two years ago by the five Founder Engineering Societies: the American Institute of Chemical Engineers, the American Society of Mechanical Engineers, the American Society of Civil Engineers, the American Institute of Mining, Metallurgical, Petroleum Engineers, and the Institute of Electrical and Electronics Engineers.

ESCOE works under a contract with the Department of Energy and all of our funding comes through that contract. Under that contract, we are to provide an independent and objective technical and engineering economic assessment activity for the Department of Energy, primarily oriented toward fossil energy technology programs.

The professional staff at ESCOE consists of approximately ten engineers in residence. Each of these residents is on loan for a two-year period from a company or, in a couple of cases, a university. These people, while they are on load to ESCOE, are one hundred percent supported by ESCOE. By design, ESCOE provides the perspective of the private sector but, because the engineers come from many individual companies, ESCOE does not have the bias of an individual company.

ESCOE is presently engaged in a number of technical tasks - about half a dozen major studies at the present time - relative to fossil energy.

The other perspective that I bring today is five years of

0-8412-0516-7/79/47-110-207$05.00/0

experience recently with the Federal Power Commission (FPC), now as the Commission's Advisor on Environmental Quality and later as the Commission's Chief Engineer and Director of the Office of Energy Systems. The Office of Energy Systems advised the Commission on the environmental, scientific, technical and economic aspects of the broad range of energy problems before the Commission.

Much of the subject of barriers to commercialization has already been discussed by the individual speakers yesterday and today. In my comments, I will make reference to where these have been expanded upon to a greater or better extent by previous speakers, and will thus try to eliminate some of the overlap.

In particular, the subject today is the barriers to commercialization of coal gasification. There will be a couple of points where it will apply to liquefaction, but the concentration will be on gasification. I will talk mostly about high Btu coal gasification (i.e., pipeline gasification) as compared with industrial gasification (i.e., low/medium Btu gasification).

As far as low/medium Btu gasification is concerned, Zeke Clark has pointed out that the barriers to commercialization there are relatively simple. In general, there is not an economic regulatory problem. There are obviously environmental problems, but, if an industry or electric utility needs the better characteristics of a gas fuel as compared with liquid or solid fuels, then industrial gasification is a viable solution. With the passage last year of the coal conversion part of the National Energy Act to greatly restrict the use of natural gas and petroleum for major facilities, there is a significant incentive for industrial gasification. There will be a significant future for industrial coal gasification as industry finds that it has few other real options when it needs clean fuel. There are no advantages to going to the additional expense of making methane to be used in an industrial process if you don't have the problem of long-distance transportation.

Now, the problem of long-distance transportation brings us to the high Btu gasification area which is our primary subject.

During the last decade there has been a rapidly growing interest in the possibility of using liquid and gaseous fuels derived from coal to partially displace conventional liquid and gaseous fuels. The interest in coal gasification has been particularly stong within the natural gas industry since the realization in the late sixties that the rate of natural gas consumption was exceeding the rate of discovery of new supplies that could be developed under prevailing federal wellhead price regulation for interstate gas. The natural gas shortages during the severe winter of 1976-77 coming on the heels of oil shortages of 1973-74 have created a broader interest in the commercialization of coal gasification within the government. However, the expectation of the last few years has not yet been translated into plants nor into products.

The fundamental barrier to the commercialization of high Btu gasification is the lack of firm government decisions to, in some manner, pay the domestic cost that is going to be necessary to reduce our dependence on foreign oil. The lack of these firm government decisions is due to confusion as to the specific barriers that must be overcome before facilities will be built and production started. These barriers to commercialization can be conveniently discussed in five categories:

- product cost,
- market insecurity,
- unproven technology,
- environmental uncertainties, and
- regulatory decisions.

The first four subjects are really an integral part of the last subject - regulatory decisions - but I will treat them in that order and come to the regulatory decision framework as the encompassing conclusion.

Howard Siegel, this morning in one of his slides, listed four areas of commercialization barriers which he then said he would not discuss to a great extent. But the four he mentioned relate very closely with these five. His first one was government approvals and regulations which is the last of the five I want to talk about. He talked about environmental clearance and identified that as a key problem. He talked about the cost and pricing of gas which is at the top of my list. He talked about the financing arrangements which is part and parcel of a couple of the subjects -- market insecurity and unproven technology -- on my list.

Of these five, I will spend most of the time on the first and the last: product cost and regulatory decisions. You have heard much about the others -- unproven technology, environmental uncertainties, and market insecurity.

PRODUCT COST

To start the discussion of product cost barrier, let me read one paragraph from a paper presented in September at a Coal Liquefaction Workshop sponsored by the International Energy agency in Munich, West Germany, where a few of us from the United States and Great Britain met with a larger group of senior technical people from a number of the major German companies that are in the coal liquefaction and coal gasification business.

> "Prior to the OPEC embargo, the general belief was that the market price of crude oil would have to about double to make coal liquefaction competitive in the United States. Five years later now, the average market price of crude oil in the United States has about tripled, but the general belief still is that the market price of crude oil must about double if coal liquefaction is to become competitive."

At that point, an engineer from Lurgi Corporation commented, "This is what we have also observed recently in Germany. We refer to it as 'chasing the receding break-even point.'" To a very real extent, that is what we are dealing with.

Figure 1 is a set of graphs showing the average U.S. field price for the three forms of fossil energy that we produce in the United States as a function of time since 1970. Inflation, which for convenience is referenced to natural gas, is also shown. Of course, energy prices themselves have contributed to inflation, but it is important that we not delude ourselves into believing that much of the problem we are dealing with is due to inflation.

Figure 1 clearly shows the effect of the decision by OPEC to increase the price of their oil. U.S. oil has tracked that increase. Since Figure 1 shows the average U.S. wellhead price, this increase in oil price does not fully reflect the world market. Some oil in this country is regulated at the wellhead.

A single point for 1978 on Figure 1 shows the higher average price paid in the U.S. when both imported and domestic oil is considered.

Another important observation from Figure 1 is the way U.S. coal price has tracked OPEC oil. Relatively, U.S. minemouth price of coal has increased more since 1970 than U.S. wellhead price of oil. Coal has tracked the OPEC price closer than it has tracked the price of our own oil because of the partial regulation of oil in this country.

And that is the major part of the problem of the "receding break-even point." Historically, the free market for energy has given minemouth coal a value of about 60-70% of the wellhead price of oil. If that ratio stays the same - and there is not much reason to assume that it would not - and given that the efficiency of converting coal to liquids is about 60-70%, liquids from coal will compete in a free energy market only when the plant

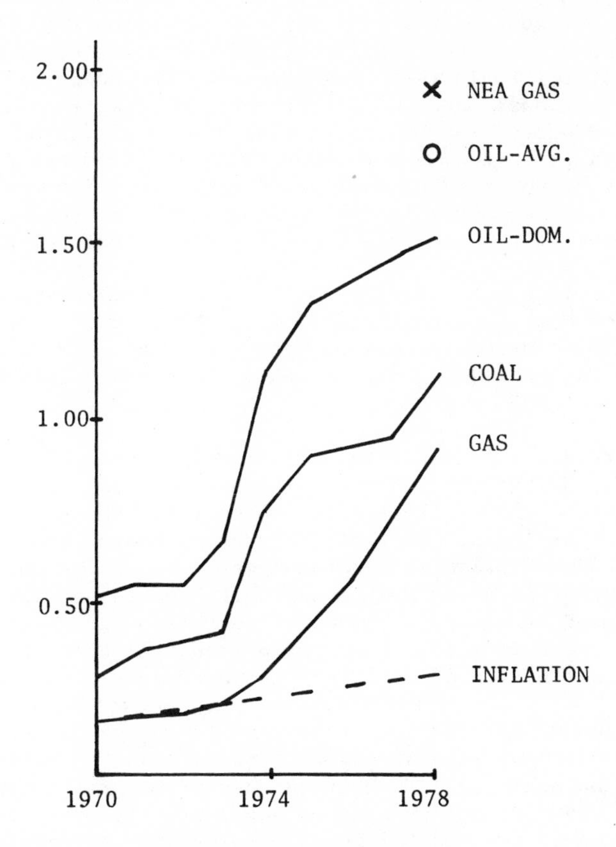

Figure 1. U.S. fossil fuel prices

cost and operating expenses are free!

Another thing to see in Figure 1 is the dramatic increase in the wellhead price of natural gas. The curve for the average price of natural gas starts at about 20¢ (per 10^6 Btu) in 1970 and is up to about 90¢ eight years later. In 1970 all interstate natural gas was being held at the low price by the stringent cost-based price regulations imposed by the Federal Power Commission (FPC) as the result of the 1954 Phillips decision from the Supreme Court and numerous other regulatory and court decisions since then. The curve starts to increase following the FPC decision in 1973 to go to a national rate and to set that national rate at about 50¢. The 1976 decision by the Federal Power Commission to raise the wellhead price from 50¢ to approximately $1.50 has continued to pull the curve up. The National Energy Act price set by the Congress, which is now a little over $2 and will escalate yearly, will continue to pull up this average price of natural gas.

The product cost barrier to the commercialization of high Btu gas from coal is significant. With the factor of four increase in the wellhead price for natural gas, production is increasing and demand is decreasing. At present, the supply of $2 gas exceeds the demand.

Howard Siegel's estimates this morning were for high Btu gas from coal at prices from $3 - $7 with his greatest confidence for the midpoint of a $5 - $7 range is three times the price for "natural" natural gas. On a strictly price basis, coal gasification is not competitive with natural gas. You may also remember that the rather infamous MOPP study that created so much furor in the Department of Energy a couple of years ago was making predictions that there is a lot of gas at prices well below the price of high Btu gasification of coal in this country.

The second barrier as far as cost is concerned is that the increasing price of natural gas will cause a lot of rethinking on the use of natural gas. Much of our use of natural gas has been built into the U.S. energy systems because of its very low price as maintained by the Federal Power Commission. Already there are significant trends away from the use of natural gas by industry and this is likely to continue. Demand for natural gas is not likely to increase in the future at the rates that were common in the 50's and 60's.

MARKET INSECURITY

The market cost barrier discussion leads directly into the subject of market insecurity and is closely related. In a completely free market, which is not the case for high Btu coal gasification, business executives must make a prediction of size and mix of the future market and then plan their facilities to meet the anticipated market. Their obvious concern is that if facilities are built to supply a certain form of energy and if another supply of the same form of energy comes in at a lower price, the product cannot be sold except at a lower price. For a regulated industry, that same question arises, but it is not so much a decision for the business as it is a decision for the regulator. In either case, someone must judge the security of the market. Just how much methane is going to be in the market in the future? Just what is going to be the impact of the Coal Conversion Act which, simply stated, requires that industries and electric utilities shall not build new facilities to burn natural gas or to burn petroleum? What is the future market for methane?

The questions of market insecurity and market price reminds me of a study the ESCOE recently completed. We were asked to do a coal fuel cycle study. The study was an examination of all the possible ways that coal from a mine could be processed and transported to deliver energy to "the city gate." After many, many pages of looking at all of the alternatives and the best estimates of price that go along with these, we were asked if it was possible to reduce the study conclusions to a single sentence. The answer is "Yes. The cheapest way to use coal is to burn it." Our expansion on the one sentence answer is "the more processing that is done, the more expensive the product."

UNPROVEN TECHNOLOGY

The next subject is unproven technology. Here a major concern is reliability. This was also referred to yesterday. The difference between a plant operating 90% of the time and 70% of the time is usually much more than the profit margin. This question is important and until a commercial size plant for a new technology is operating, the answer is uncertain.

Another major concern with unproven technology is the capital cost. In the study of coal liquefaction costs that I referred to earlier, we saw plant capital cost estimates which increased by a factor of three since 1970. Howard Siegel referred to this in his comments when he was politely pointing out that some people's estimates were not as good as some other people's estimates. Demonstrations of commercial size technology are needed to get a better handle on reliability and a better handle on the actual cost.

ENVIRONMENTAL UNCERTAINTIES

Environmental uncertainties have been well handled by previous speakers. My only comment relates to Howard's remark that as we build more coal-refining plants, we can expect the price to come down something like 10% or 20% because of experience. He then pointed out that there were some environmental questions as to just exactly what the requirements for effluent control are going to be in the future. I will give you an off-the-cuff estimate that the cost of the answers to those environmental questions will more than offset reductions in price as we go on. There will be more stringent environmental requirements, and this will boost the cost if and when a coal refining industry develops.

REGULATORY DECISIONS

Figure 2 is a diagram of the organization of the Department of Energy. Of particular interest is the Federal Energy Regulatory Commission which is a "part of the Department of Energy."

Under the Reorganization Act for the Department of Energy, the agency is directed by the triumvirate of the Secretary, Deputy Secretary and Under Secretary.

The six boxes on the right, under the Under Secretary, are responsible for the DOE outlay programs. These are the programs which support R, D&D and manage some of the physical energy resources and plants owned by the Federal Government. Ninety-six percent of the Department of Energy budget goes to these outlay programs, including the commercialization programs that Dick Passman was talking about this morning. The Assistant Secretary for Resource Applications has the responsibility for the Fossil Energy Commercialization Program.

On the left side of the diagram are the information, policy and regulatory activities which are the responsibility of the Deputy Secretary. The Assistant Secretary for Policy and Evaluation is responsible for all of the studies leading to proposed legislation. This is where, for example, the National Energy Act was developed in detail. The Economic Regulatory Administration has the responsibility for regulations such as conversion of power plants to coal and oil price regulation - all regulatory authority not in FERC.

The Federal Energy Regulatory Commission (FERC) is not significantly different from what the Federal Power Commission was previously. The Federal Power Commission was an independent regulatory agency and FERC is an independent regulatory agency. In Figure 2 there is no direct line from the Secretary to FERC. The five Commissioners are appointed by The President, not by the Secretary, and must be confirmed by the Senate. They cannot be relieved of their duties during their four-year terms except by impeachment. The DOE Organization Act forbids the Secretary from directing the activities of FERC in any manner. In fact, the law specifies some decisions that the Secretary can make only

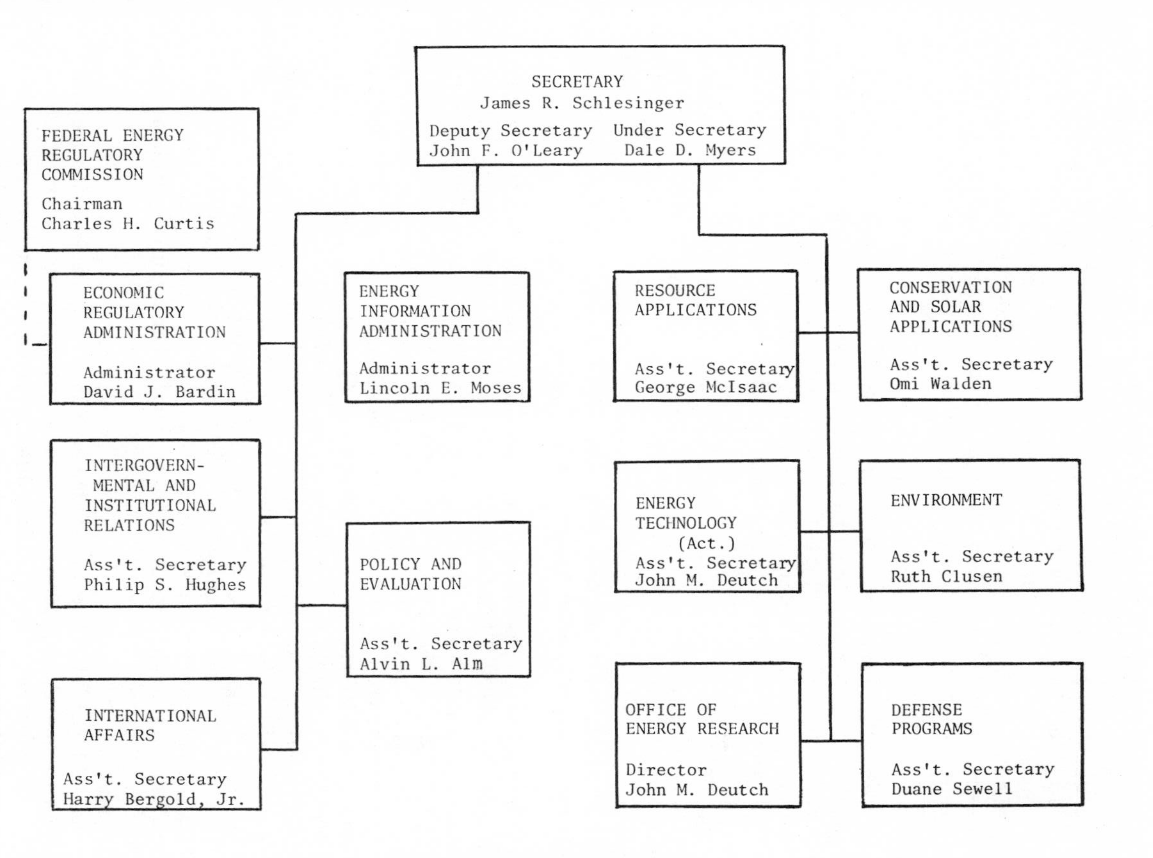

Figure 2. DOE organization

if he has the approval of FERC. Under the law, the Secretary is allowed automatically to be a party to any case that he chooses before the Commission - which is exactly the same right that is given to all State Public Service Commissions. The Secretary has delegated the responsibility for appearing before FERC to the Economic Regulatory Administration - thus the dotted line with an arrow going to FERC in Figure 2. However, under the law, they are an equal party with everybody else in all decisions before the Commission. Thus, for example, in the Great Plains Gasification Case, the opinion of the Secretary of Energy can carry no more weight in FERC decisions than any other party to the decision at hand.

It is also important to understand the distinction between the staff of the Commission and the Commission itself. When it is reported that the FERC staff has taken a certain position, many people interpret that as a Commission decision. That is not a correct interpretation. The FERC staff, by law, is independent; it has an obligation to protect the public interest as they see it. The FERC staff appears before the Administrative Law Judge in all hearings as an equal party. The position of the staff is not the position of the Commission unless the Commission later adopts it. In fact, the Commission has a record of going against the staff about half of the time and for the staff about half of the time; about the same ratio for any other party.

For high Btu gasification, the question is: What will the Commission decide? First of all, it is clear that before high Btu gas can be produced in a coal gasification facility and introduced into an interstate pipeline system, the Commission must approve the transport and sale of the product. And the basic test that the Commission must apply is: Is that cost just and reasonable and is it in the public interest? The basic question must be: Are there cheaper ways to assure an adequate long-term supply for the consumer? That is a basic difficulty that FERC is going to have and did have back in the early seventies when it was dealing with the Wesco and the Burnham gasification applications that were then before the Commission. The Commission must find justification, that can be defended in the courts, for consumers paying a price that is significantly above that which would purchase gas from other sources.

An FPC rule-making about three years ago resulted in Order 566 wherein the Commission made a specific change in its rules to allow a consortium, or an individual company, to treat as an R&D expense a portion of the expense of a commercial size demonstration plant for new technology. This course was not what was chosed by Great Plains, but the suggestion has been made by some FERC staff that this may be a way to justify the difference.

It is not important whether Great Plains adheres to the letter of Order 566. It is important, however, that the Commission adopt the philosophy that was inherent in Order 566 and recognize the social value in the additional expense necessary to demonstrate a technology with important future value.

I hope that the Commission's decision will be favorable because, although I would argue that high·Btu gas from coal is not now economically competitive, we must proceed in order to reduce the technical and economic uncertainties. We must move on with the construction of one, two or a very small number of commercial-size plants so that we can learn about the real economics, and the real technical and reliability problems.

Hopefully, we will proceed in spite of these commercial barriers.

RECEIVED July 2, 1979.

Energy and Society

HENRY R. LINDEN

Illinois Institute of Technology, IIT Center, Chicago, IL 60616

The impact of the substitution of inanimate energy forms for human labor on social, economic and political developments is examined. The relationship between energy abundance and affluence, egalitarianism and physical mobility is considered in the light of the widely debated premise that it is the fundamental basis for all social progress. The growing aversion of the main beneficiaries of energy abundance and high technology - the urban intelligentsia of the Western world - to these basic sources of their disproportionate political and cultural influence is also examined. The most recent manifestation of this phenomenon - the attack on energy-intensive lifestyles and on the complex and centralized systems needed to bring the benefits of energy abundancy and high technology to the broadest possible segments of the public - is given special attention. Finally, a brief look is taken at the special responsibilities and opportunities of the United States in facilitating the critical and potentially dangerous transition from exhaustible to inexhaustible energy sources.

Energy has become a major public policy issue equivalent in importance to social, economic or defense policy only in relatively recent times. It is true that most industrialized countries had energy ministries for some time (except for the United States, where cabinet status was given to energy only in 1977), and even in countries with large private sectors, government involvement in energy supply and use has always been substantial. Government regulation, control or even ownership has been traditional for public utilities providing electric or fuel gas service, and control of motor fuel prices through taxation has been the practice for many years. The national security and economic implications of reliable sources of critical energy materials also

0-8412-0516-7/79/47-110-219$05.00/0

have been well understood, especially by the oil-poor major powers involved directly or indirectly in World Wars I and II. The socialist countries were probably the first to recognize the ideological factors in energy policy. In spite of this long history of government involvement, energy has emerged as a major idelological and philosophical issue in the Western world only in recent years.

The realities of the world energy situation pose a serious dilemma. On the one hand, the unprecedented high level of social and economic well-being of much of the industrialized world is unquestionably due to the increasing substitution of inanimate sources of energy for human and animal labor. The recent progress toward the full emancipation of women and minority groups in the industrialized world is merely the latest manifestation of a process which has gathered momentum since the abolition of slavery, serfdom and child labor. The pressure to achieve a still more affluent, mobile and egalitarian society is strong, as is the pressure in the less developed countries to achieve a quality of life more closely approaching that of the industrialized world.

On the other hand, there is growing concern that this unprecedented rate of progress since the advent of the scientific and industrial revolutions may be a transient phenomenon. Clearly, it is rapidly depleting the stock of readily available inanimate energy sources stored over long periods of geologic time, without any assurance that a new generation of practical energy supply and utilization systems will be ready when needed. Thus, man's liberation from having to depend primarily on energy and raw material sources renewed quickly by the sun may be short lived, unless the promise of new technology capable of utilizing the less readily available energy sources can be realized in a timely fashion.

The most worrisome aspects of this rapid social and economic progress is that it has been accompanied by a tremendous increase in world population growth - from an annual rate of no more than one tenth of one per cent prior to 1750, to a recent peak of about two percent. This rate, although finally declining, still continues at a pace which inevitably will yield a world population of more than eight billion well within the first half of the 21st century. This would be at about the same time when the availability of the energy sources that can be utilized most easily - first crude oil and later also natural gas - will be severely restricted. Thus, in the absence of options that would sustain these and even higher world population levels under conditions conductive to social and economic stability, the way down in quality of life may be as steep as the way up. In this connection, it is sobering to note that the primitive solar economy which existed from the dawn of civilized man until the fossil fuel era never supported more than a billion people, and that in a submarginal way at best.

One positive aspect of the current situation is that there is still time to develop alternatives to fossil fuels. According to current assessments, total remaining recoverable world oil, natural gas, oil shale, tar sand, and coal reserves and resources give us about 100 years of lead time at a primary energy demand growth rate somewhere between 2 and 3 percent annually. (The recoverable uranium resources, if used in burner reactors, would not extend this time significantly.) This assumes, of course, that there will be institutions with the huge capital resources and the managerial and technical capabilities needed to find, produce, process and market these large remaining fossil fuel reserves and resources. Implied also is a condition of free world trade in energy materials and technologies so that they can be shared in an equitable fashion. These are extremely large assumptions.

Thus, energy policy makers face an exceptionally severe challenge. They must find politically acceptable ways to produce and market the remaining oil and gas resources in quantities and at prices which do not impair the capabilities of the industrialized world to manage the transition to inexhaustible energy forms. They must determine the path of the transition: Is it to rely primarily on the still abundant coal, bitumen and marginal hydrocarbon resources in conjunction with synthetic fuels and a moderate increase in electrification? Or, is it to follow a high electrification scenario based on coal and/or nuclear fission? Or, must they assume that a coal - or fission-based transition is not feasible for environmental or political reasons, so that they must jump directly to inexhaustible energy forms, most of which are still far from technical, economic and environmental validation, i.e., fusion and the various direct and indirect solar options including solar thermal, photovoltaics, wind, hydro, ocean thermal, biomass, etc.? In choosing any of these paths one thing is certain: It will require a substantially higher share of economic output for the energy sector than during the golden age of abundant and cheap crude oil and natural gas.

Differences in National Energy Policies

In the United States, the traumatic realization that energy self-sufficiency had been lost led to the first Presidential pronouncement of an overall energy policy in 1971. It advocated programs to increase the development of domestic hydrocarbon resources, to use more coal in environmentally acceptable ways, to develop synthetic substitutes for crude oil and natural gas, and to provide more electricity by nuclear fission. The top priority for Federal support was the liquid metal fast breeder reactor. Overall, the emphasis was on more domestic supply and an assured energy future. Of course, some adjustments in priorities were made during and following the 1973-74 oil embargo in recognition of the need for more strenuous measures to keep a lid on oil imports. Growing pressure for energy conservation and use of solar and geothermal energy also was accommodated by the Nixon and Ford Administrations. However, the main thrust was to increase domestic supplies on behalf of energy self-sufficiency, or "independence," and to do so with whatever pragmatic solutions were available. These policies implicitly recognized the great contributions that cheap and abundant energy supplies had made to society. They were designed - however imperfectly - to continue these contributions for the benefit of future generations.

An interesting aspect of this policy was that protection of U.S. military capabilities was seldom used explicitly for its justification. The relative complacency of the Western European countries and Japan about their much greater lack of "energy independence" contrasted sharply with the growing concern about this issue in the United States. The obvious difference is, of course, that the United States has the additional role of military protector. This role is much more difficult to play when a major portion of as critical a strategic commodity as petroleum has to be imported from sources offering dubious security of supply. Other, more widely used but, at the time, ineffective arguments in support of "energy independence" included:

1) The need to make foreign policy decisions unencumbered by the threat of another oil embargo.
2) The need to maintain an acceptable balance of trade.
3) The prospect of early depletion of world crude oil reserves with accompanying rapid escalation of world oil prices, so that growing dependence on oil imports would not, in any event, be practical.
4) The immorality of depriving less developed countries, not as capable as the United States in meeting their needs with domestic energy supplies, of the oil that is available on the world market.

Recently, because of the sagging U.S. dollar, the balance of trade argument for increased energy self-sufficiency has become more effective and has become an increasingly pervasive issue affecting energy policy. It is, of course, true that the United States cannot match the ability of West Germany and Japan to offset energy imports with exports of manufactured goods. However, there is a question to what extent the increase in the large U.S. trade deficit caused by the purchase of $45 billion of foreign oil, and to what extent U.S. monetary and economic policy in general has been to blame for the decline in the dollar.

The changes in U.S. policy direction proposed by President Carter in April 1977 were not so much substantive as philosophical. In fact, with the exception of the deferral of the breeder reactor, the specific initiatives of President Carter's energy plan were very similar to those proposed under the Ford and Nixon Administrations. What changed was that continued increases in energy use to improve the human condition were no longer portrayed as something desirable, to be compromised only temporarily under the pressure of national security and monetary problems. The ensuing national debate about the severity, means for achievement and likely consequences of the original energy plan's proposed limits on energy use was responsible for the many modifications incorporated in the 1978 National Energy Act and for the growing and politically very healthy consensus on what constitutes an appropriate energy policy for the United States. The extremely critical proliferation issue, which has indefinitely deferred the development of a U.S. breeder reactor, unfortunately was not resolved in this debate. Without the breeder, the contribution of nuclear energy to total energy supply would be marginal indeed. With it, the world would have assurance of long-term energy abundance, admittedly at considerable cost and environmental and security risks. However, prospects for the breeder are dimming in view of growing evidence that anti-nuclear sentiment and, in particular, anti-plutonium economy sentiment is more pervasive than evidenced by the earlier U.S. referenda, and also spreading in Europe.

The attitude toward energy policy in much of Western Europe and in Japan provides an interesting contrast to U.S. policy. The philosophical constraints that have hampered U.S. energy policy in recent years appear to be much weaker there. Consequently, government intervention into the energy market seems to be more pragmatic and consistent with other national policy objectives than in the United States, where the most productive balance between free market forces, regulation and state ownership remains a subject of great controversy. Also, the conservation issue has not assumed as ideological a character in Western Europe and Japan as it has in the United States. Conservation outside of the United States is accepted as a normal response to energy pricing policies which reflect the realities of the energy supply situation and of an infrastructure of energy consumption developed over a long history of relatively high energy costs and questionable security of supply.

Energy and Political and Social Philosophy

There are currently four clearly distinguishable positions on the relationships between energy and society put forth by various interest groups which I shall call the economic purists, the intellectual elitists, the technocrats, and the materialists.

Let me stress at the outset that my intent here is to classify, not to endorse or criticize.

The economic purists, mostly academicians of conservative leanings moving freely between universities, think tanks, research institutes and government, hold the view that energy must be treated like any other economic good - infinitely substitutable at high elasticities by capital investment or labor. A corollary of this view is that it is unlikely that there will be discontinuities or drastic changes in the slopes of the supply curves for fossil fuels and uranium oxide -- i.e., relatively moderate increases in price will always yield relatively moderate increases in supply, and such higher prices will reduce demand sufficiently and rapidly enough to bring it in balance with supply without any major economic or social disruptions. Further, this view holds that if one source is interrupted by human intervention or some other externality, there will always be another to take its place. The economic purists refuse to assign energy any special role in terms of its impact on society and tend to treat any warnings about an impending "energy crisis" with disdain. This view tends to have history on its side - a widely predicted crisis has seldom materialized with anywhere near its projected severity. And, they ask, what kind of a "crisis" is it which is now in its sixth year without having caused any permanent dislocation apparent to the general public? However, this sanguine view of energy does seem a bit risky in light of considerable evidence that:

1) In many respects energy availability in association with energy utilization technology has been a more important tool of social progress than advances in religious, philosophical, or political thought.
2) In a strictly material sense, energy is a lot more like food than any other economic good, so that an energy famine in a society heavily dependent on energy for economic and social stability would have effects similar to a real famine and, thus, the consequences and potential remedies simply cannot be put in strict economic terms.
3) In addition, and very importantly, energy has much greater strategic value that most critical commodities, such as steel, and in this sense the need for energy can be equated to the need for high-technology weapons and the capability of producing, deploying and using them. One important strategic property of energy in the form of fossil fuels is that the sheer magnitude of the quantities required and their unit storage costs are such that stockpiling of fossil fuels is generally much more difficult and costly than stockpiling of other strategic materials.

A second view of energy and society which has become extremely fashionable is that of the major segment of the intellectual elite of Western society. This elite can be broadly characterized as liberal or left-leaning, but not Marxist; nearly entirely urban; generally idealistic and relatively young; and influential beyond its numbers because of its concentration in the press and other media, in education, and in government. The views of this group insofar as energy and society are concerned are colored by their antagonism toward both the international oil companies and OPEC, a distrust of free market economics, and more than a tinge of romanticism. This romanticism is evidenced by a somewhat superficial show of antimaterialism expressed in a dress code emulating that of farmers and blue collar workers, an aversion to conventional automobiles, and a predilection for "soft" (i.e., low technology) solutions to the energy problem. This, in spite of their affluence and their total dependence on energy-intensive high technology - i.e., jet planes, computers, communication satellites, television, etc., for their status, influence and mobility. The intellectual-elitist position on energy and society stresses measures which would impact most heavily on others; i.e., conservation for those who are not affluent enough to waste, sacrifice for those who have little to sacrifice, and mass transportation for those who have just escaped its rigors. Another fundamental characteristic of the intellectual-elitist view is its strong anti-nuclear bent and its hostility toward all centralized systems of energy supply and utilization because of their supposed inefficiency and ability to withstand local control by the consumers. Dispersed, small systems are considered superior a priori as exemplified by my good friend Amory Lovins' doctrine of the "soft" vis-a-vis the "hard" path.

A corollary view held by the intellectual elitists is that the United States is an energy wastrel when compared with other industrialized nations such as Sweden, West Germany, and Japan. It is argued that these countries have shown a lower ratio of energy use to Gross Domestic Product (GDP) or Gross National Product (GNP) and lower per capita energy consumption while maintaining the same standard of living as the United States. I will discuss this in more detail later; suffice it to say now that such comparisons can be misleading for a number of reasons, including the fact that they neglect such real indicators of a nation's prosperity as its citizens' purchasing power.

A third and very influential view of energy and society is that of the technocrats. They are found largely in the executive branches of central governments, in government research organizations, and in other institutions closely allied with government. To them, the energy problem presents an opportunity to increase their influence over energy policy through massive and often redundant studies followed or accompanied by the imposition of complex new regulations, controls and taxes. They have important allies among those who see the energy problem as an opportunity for attacks on the private sector and for new social engineering and income redistribution schemes.

Moreover, the energy problem gives technocrats both inside and outside of government the opportunity to administer and implement greatly expanded R&D efforts and associated research in the soft sciences. The technocratic model of the energy problem projects impending disaster due to the inadequacy of world petroleum supplies and other essential resources. This crisis atmosphere tends to polarize support for the various technological options. For example, those advocating substitution of coal or coal- or oil-shale derived synthetics are opposed by environmentalists, ecologists and advocates of nuclear and inexhaustible energy forms because of a wide variety of disastrous consequences predicted for greater fossil fuel use. The latest and most tenuous of these predictions is the "CO_2 catastrophe." Nuclear energy, which seemed to be the consensus solution to all energy problems only a few short years ago, has come under the most concentrated attack by a broad coalition of diverse interests. It includes not only the traditional opponents of fission and fusion, but also a large faction that has switched its support to solar energy and its derivatives. Meanwhile, the confrontation continues between advocates of an all-electric economy and those who want to preserve the present infrastructure based on fluid chemical fuels.

The remaining view of energy and society which can be put in a single category is found among surpisingly diverse elements of society who can be broadly characterized as materialists. These elements include traditional Marxists, the managerial class in private industry, conservatives from many social and political strata and the majority of the non-ideological, non-political labor movement found primarily in the United States. These diverse groups view energy as the engine of economic progress and of upward social and economic mobility. The political conservatives also view energy abundance as an important source of political freedom and freedom from government interference in predominantly materialistic outlook of traditional Marxism which, for non-ideological reasons, is shared by much of labor and business. The obvious interrelationship between the substitution of energy and capital-intensive devices for human labor and the elimination of economic exploitation is apparent to these groups dedicated to raising the general standard of living. The Marxists, however, have to face the equally obvious interrelationship between increasing energy abundance and increasing social mobility, which is generally accompanied by a desire for greater political and intellectual freedom. By traditional Marxists, I mean those in control in the Soviet Union, the Peoples' Republic of China, and Eastern Europe. By contrast, the intellectual leftists of Western society are more closely allied to the liberal elitists. Thus, they tend to take an anti-energy, anti-technology stance, perhaps because they, too, feel threatened by increased social mobility. Because the leadership of many disadvantaged groups in Western society comes from this intellectual elite, certain conflicts exist between the interests of these disadvantaged groups and the ideology of its leadership. For example, some leaders of the women's rights movement profess anti-energy and anti-technology ideologies because they fail or do not wish to recognize the linkage between the emancipation of women and the substitution of inanimate energy for the cheapest and most abundant source of human labor - the exploitation of the wife and daughter by the dominant male head of the family. In refreshing contrast, the leadership of the most prestigious U.S. organization defending the rights of blacks, the National Association for the Advancement of Colored People (NAACP), has endorsed a pro-energy, pro-technology development stand in recognition of the obvious self-interest of its constituency.

National Energy Consumption Patterns and Energy "Waste"

My own position is still evolving, but I have fully accepted the overwhelming evidence that a society's economic and social well-being is directly linked to its use of energy. In most industrialized nations, real income has risen or fallen in unison with per capita energy consumption. Further, primary energy demand has been relatively price inelastic, although this is a subject of great debate among energy modelers. In the United States, a ten percent increase in the deflated price for fuels and power has reduced primary energy consumption only by about two percent, and vice versa, according to the admittedly simplistic analyses performed by me and my associates. Moreover, when we look at the relationship between per capita GNP or GDP and per capita energy consumption for industrialized nations, we see that no country has been able to increase the production of goods and services without the expenditure of an additional amount of energy and that, indeed, the amount required to do this has been roughly comparable in recent years.

I am also concerned with the philosophical validity of restricting energy consumption by labeling certain uses as "waste" The definition of "waste" is clearly based on very subjective value judgments conditioned by ideological and cultural preferences. Man simply is not an "efficient" being. He is a ceremonial creature who employs his tool-making and tool-using capabilities to a considerable extent for conspicuous display. In fact, what distinguishes man most clearly from all other species is that he spends a major portion of his energies and resources on concerns other than survival, such as building monuments and acquiring symbols of power and status. Therefore, man's most important achievements in the eyes of history would seldom pass the test of energy efficiency. This includes Stonehenge; the Egyptian, Mayan, and Aztec pyramids; most of the edifices identifying seats of political, social and economic power since the beginning of civilization; all of the temples, cathedrals, pagodas and mosques; and, more recently, the space program. Add to this circuses, fireworks, bonfires, torchlight parades, spectator sports, open fireplaces in centrally-heated houses, and just about any human activity that seems to have historical, esthetic or spiritual value. The philosophical basis for singling out energy as a target for attack on "waste" is, therefore, weak. There are ample pragmatic reasons, of course, under today's conditions, but this is not the issue. Restriction of energy use for philosophical and ideological reasons is the issue, especially in the face of what are clearly abundant resources of a variety of energy forms.

Nevertheless, in the United States, the question is increasingly asked: What about Sweden or West Germany, whose per capita energy use is substantially below that of the United States, while GNP's or GDP's per capita are roughly comparable? This is taken by many as evidence of American wastefulness and used to justify an energy policy based primarily on conservation rather than increased supply. However, serious weaknesses in the methodologies used in such comparisons cast doubt upon the conclusions and the policy decisions derived from them. For one thing, no model has yet been developed that takes adequately into account the effects of the size of a country, population density, climate, degree of industrialization, the mix and energy intensiveness of industry, energy prices, state of technology, the age of manufacturing equipment, historical standards of living, and other important variables. Severe difficulties are encountered in converting GNP or GDP values from national currencies to a common monetary unit, which also affects the validity of the results..

Consequently, it is misleading and perhaps even meaningless, to evaluate a nation's standard of living on the basis of per capita GNP or GDP. A better approach would be to compare the purchasing power of citizens expressed in terms of the number of hours they must work to buy a representative market basket of goods and services. A recent study based on the necessary data for May and early June 1976 by the Union Bank of Switzerland revealed that the purchasing power of workers in six North American cities - four in the United States and two in Canada - is higher than that of workers in 38 other cities in the world because of a combination of relatively high wages and relatively moderate prices. As a result, U.S. workers in selected occupations could buy the market basket of goods and services with the gross earnings from 66 to 77 working hours and Canadian workers with the gross earnings from 83 to 85 hours, whereas in Zurich, 92 hours were needed to buy the same basket, in Dusseldorf 100 hours, in Stockholm 104 hours, in London 124 hours, and in Tokyo 162 hours. When net earnings were compared, that is, salaries and wages after deduction of taxes and social service contributions, the results did not change materially. This lends credence to the view that the United States' and Canada's relatively high energy consumption may be related to the high purchasing power of its citizens. There is, I might add, some evidence that overinvestment in energy conservation has hurt the Swedish economy. In recent years, Sweden has suffered a large drop in capital investment, rising unemployment and a very high decline in real GNP. Other factors undoubtedly contributed, but this may be an indication that an industrial society cannot restrict its energy consumption unduly. In fact, I suspect that the undesirable structural changes in the world economy following the 1973-74 oil embargo (i.e., lower real economic growth rates, higher levels of unemployment and higher inflation rates) may, in part, be a consequence of the general reluctance to increase energy use. Even at today's high prices, substitution of energy for capital and labor may still pay off in many instances.

I do not want to imply that significant improvements in energy use could not be made that are both practical and beneficial. In fact, considerable progress has been made and is being made in implementing such improvements. The fuel consumption of the notorious American "gas guzzling" automobile will reach 27.5 miles per U.S. gallon under Federally mandated standards by 1985. U.S. industry, in particular, has made significant strides in improving its energy efficiency, indicated by a steady decline in the overall ratio of energy use to GNP since 1970 to an all-time low in 1978. However, even under President Carter's April 1977 energy plan, with its emphasis on conservation, U.S. oil imports were projected to be 5.8 to 7.0 million bbl/day by 1985 - 34 to 38 percent of total consumption. Actual U.S. oil import requirements are likely to be very much higher, (more than 9 million bbl/day in 1979, and probably 12 million bbl/day in 1985) unless a prolonged economic recession occurs. The open question is, of course, if these quantities will, indeed, be available. The good news from Mexico was quickly offset by bad news from Iran.

U.S. Responsibilities in Stabilizing the World Energy Situation

Excessive U.S. dependence on world oil supplies could have disastrous consequences. As world crude oil production approaches its peak - certainly not later than 2000 to 2010 - precipitous price increases to a level equivalent to the replacement cost of liquid motor fuels by synthetics would occur. We are talking here about at least $30/bbl and, more likely, $40 (in 1978 dollars) The burden of this would fall most heavily on Japan and Western

Europe, and the less developed countries without significant oil and gas resources. This, in turn, could result in economic depression, internal political instability and, possibly, armed conflicts (i.e., resource wars) - all of which would place an enormous financial and military burden on the United States. Certainly the U.S.S.R. and its primary trading partners, in spite of what appears to be a somewhat more favorable domestic energy resource picture, would not want a serious world crisis over energy in view of the sad history of previous world confrontations over essential raw materials.

Thus, it would be in the self-interest of everyone if the United States could reduce, or even eliminate, its dependence on imported oil. However, energy autarky, while beguiling from the national security viewpoint, is not practical economic policy for the United States, or anyone else for that matter. The concept of energy independence has recently given way to the more rational concept of a hierarchy of oil (and gas) sources ranked in accordance with their cost, resource potential, security of supply, environmental and economic impact, and impact on monetary stability. Clearly, in such a hierarchy, Canadian and Mexican hydrocarbon sources rank very high, as do certain other sources of imports. Within this expanded concept of energy independence a satisfactory level of security of supply at acceptable economic, social and environmental costs would be attainable by the turn of the century if the United States reverses its de facto policy of interminable delay of full development of its domestic energy supplies. The foundations of such a new policy would, on the basis of any realistic assessment of the options, have to be accelerated exploration, development and utilization of the vast remaining conventional and unconventional natural gas resources and creation of a large synthetic fuels industry based on the abundant U.S. coal and oil shale resources and on the utilization of biomass materials where this makes economic sense. The logical counterpart to such an enhanced domestic supply policy would be to reserve liquid fuels, both natural and synthetic, for transport uses where they have maximum form value and to put gas, both natural and synthetic, back into all of its traditional stationary heat ;energy markets action by any other major industrial power or group of powers would contribute more to stabilizing the world energy situation and to easing of the transition to inexhaustible energy sources.

Conclusions

What, then, are the conclusions from this review of the role of energy in society? Clearly, the conclusion that all industrialized countries have to pursue <u>every</u> option to assure future energy abundance is too simplistic. Even if there were no physical limitations to such an approach, political and economic realities alone would dictate the setting of priorities. Yet, heroic measures are needed to reverse the lemming-like march of the industrialized world toward the imminent point when world oil productive capacity will peak and, simultaneously, the less-developed and much more populous countries will clamor for an increased share of this diminishing essential resource.

No degree of politically tenable self-denial by the industrialized world will provide for the rapidly growing energy needs of the world. These will only be satisfied through the full utilization of all economically and environmentally acceptable energy sources in a climate of free world trade, combined with an extremely intensive effort of research, development, demonstration and commercial deployment of new energy technologies by the industrialized countries at budgets similar to what is now invested in defense. Critical to accomplishing the transition from exhaustible to inexhaustible energy forms, within the existing time constraints, is the full development of the world's huge coal, bitumen and marginal hydrocarbon resources and of the means for their utilization in economically and environmentally acceptable ways.

In exploring alternatives to nuclear fission and fusion, it is particularly important to determine as quickly as possible whether a combination of solar and biomass options can provide the food, fiber, shelter, transport and other essentials for a world population that could easily reach 10 billion well before the end of the 21st century. We know enough today to explore within reasonable limits of certainty whether a totally non-nuclear economy in the post-fossil fuel era, be it high-technology or low-technology, can provide the necessities of life for this number of people.

In case we fail to provide technological solutions for meeting the world's growing energy needs, we must face the ultimate reality. In the past, strong nations have always reacted to shortages of critical commodities through war and imperialism while the weak nations bled and starved. This is an option that, aside from its immorality, does not seem very practical at a time when nuclear weapons have proliferated widely. Therefore, full utilization and equitable exchange of the world's technology and energy resources is essential to an acceptable future for mankind.

RECEIVED May 21, 1979.

15

Roundtable Discussions

R. WOLK: Since we had papers from Exxon both on liquids and SNG, could Howard Siegel give us some idea of the relative cost of SNG and liquids from eastern and western coals?

PANELIST SIEGEL: The relative cost of gas and liquids from coal depends a great deal on the coal that is being liquefied or gasified. For example, with Illinois coal, our information would indicate that it is quite readily liquefied but is very difficult to gasify, and the cost of SNG or IBG from Illinois coal would be higher than the cost of coal liquids. It might be as much as 15-20% higher. But by changing, for example, to a Wyoming location and a surface-mined Wyoming type coal, liquefaction becomes more difficult than with Illinois coal but gasification becomes easier. The western coals are very reactive to gasification and they cost less per ton. Compared with Illinois liquids, the cost of gas from the western coal would be lower than the cost of the Illinois liquids. It might be 15-20% lower. In a sense, we have a spread where gas varies from perhaps 15-20% higher to 15-20% lower than the cost of liquids depending on the coal-feed to gasification.

L. E. SWABB, JR., Vice President, Exxon Research & Engineering Company: A question for Dick Hill.

On your chart showing the cost versus time of oil, coal and gas, you had the coal rising somewhat in parallel to the price of oil, and the implication was that the coal was really responding to the rising cost of oil. It seems to me that it is probably more complicated than that, and I wonder if you have made any analysis of the effect of the various laws, the Mine Safety Act, etc., on the price of the coal.

PANELIST HILL: The answer is "No." We have not attempted any significant analysis. One of our people was slightly involved in the question of the impact of the new surface mining laws. Of course, that would be a future cost, not part of the costs that were shown there.

I would agree that the question of the price of coal is a rather complicated one. But what the curves and I think we sensed at the time is that basically the value of coal is very determinant

0-8412-0516-7/79/47-110-233$05.00/0

on the price that people are willing to sell it for, and that as the price of oil increases, the free market will pay a higher price for coal. People who have a vested interest in coal in the ground are receiving a greater price for that coal. Indeed there is a lot of expense buried in coal prices. But the manner in which coal prices increased sharply at the time that the OPEC prices of oil increased and the fact that they have stayed reasonably in step demonstrate that basically there is a free competitive market between coal and oil. Coal and oil today is probably more directly competitive in the generation of electricity than it is anywhere else. Electric power generation uses such large quantities of oil and coal. There is some exchangeability between the two and they track along very nicely.

I would contend that indeed additional legislation will alter prices, but that you will find that domestic coal will track the value of energy quite well.

The natural gas price is not reacting to what you might consider a free-market situation. I contend that it will eventually if the amendments to the Natural Gas Act that were part of the National Energy Act last year are not altered significantly. There will be a relatively free market and soon the control price will become a ceiling and the market will be trading under that ceiling. But at the moment, that sharp, almost exponential curve is a natural response to the release from the ridiculously low wellhead price control that has been put on natural gas by the Federal Power Commission following the Phillips decision of 1954.

GENERAL CHAIRMAN PELOFSKY: A question for Dick Passman.

You, mentioned many coal programs in your talk this morning. Have you prioritized them in some order? Are you giving more emphasis to one over another? If so, which one?

PANELIST PASSMAN: Not really in terms of the various kinds of supplemental fuels. Low and medium could be considered almost together. We are really looking for the industry to say what their applications are and state their willingness to do studies because they intend, when the conditions are right, to put in a plant. We hope to get a variety of responses from industries such as chemical feedstock, different regions, types of coal, and provide some measure of support for this feasibility and planning study. We want to do that as soon as possible.

On high-Btu gas, we are looking to see if there really is an interest. If there are contracts for a large supply of Canadian and Mexican gas, even if there is a favorable treatment by FERC of the Great Plains Gasification Project, I think we really need some personal response from the major companies that have been pursuing the pipeline gas to gain a better appreciation of what kind of support is needed. I think you can tell from my talk that I consider this to be a very important resource. High-Btu gas fills almost every market niche. Only part of industry could be supplied by low- or medium-Btu gas. But there are many people who can't have coal piles in their backyard and may operate a oneshift operation in a city location. They turn on the gas in the morning, use the energy for their work, and turn it off at night.

So I, personally, am very supportive. But if we are to commercialize something, and the major supplies of pipeline gas are really not interested and for good practical reasons, I think we have to know what those reasons are.

We set for ourselves about a four-month limit to make our evaluation and determine how we proceed from there. It is a very high priority item. We are going to do the study first. We are not going to guess any answers.

Concerning liquids, the priority in our shop lies with the indirect processes, which are currently commercial to find out what the economics really are in building a plant under the conditions in this country so that a better evaluation can be made. It has to be compared with the direct processes which basically are aimed at backing out oil through providing a process capapability, whereas, I think the real urgency is the transportation market, particularly if people are upset by lines at the gas pumps which are now possible. So I indicate that I don't agree with the schoolbook approach where the demand, supply curves come together gradually. I think that it's major events such as embargo, overthrow of a country and so on that cause sudden changes and have to be protected against.

One of our obligations in the Department of Energy is to see that we in the country can handle it. I see the capability that the top priority in the liquids area would be transportation fuels.

B. SCHMID: I have a question for Howard Siegel.

In the gasification process, do you feed the coal as a slurry or as a solid?

PANELIST SIEGEL: The coal is fed as a dry solid. The steps are, first, to grind and crush the coal, then to dry it, then to impregnate it with a water solution containing the potassium catalyst, then to dry it again and feed it as a dry solid to the gasification bed through a lock hopper arrangement with gas injection into the bed.

B. SCHMID: And one other question on the yields. In addition to gas, do you get some liquids, and also how much carbon is associated with the ash when it is rejected?

PANELIST SIEGEL: The program that we are working on with the Department of Energy is aimed at a process to produce SNG with no by-product liquids. The way in which we arrange for that to happen is to feed the coal below the top of the bed--feed it near the bottom of the bed--and then all the liquids are destroyed, gasified in the bed along with the coal, and the only product is gas.

B. SCHMID: The latter part of the question was: How much carbon would be rejected, if any, with the ash?

PANELIST SIEGEL: We visualize the ash removal will be done in two ways. One way is that the fines that come overhead will, based on our past experience, be higher in ash content than the material in the bed. So we will withdraw a portion of the ash that way, and withdraw the rest of the ash by taking a purge stream out of the fluid bed. Roughly, the withdrawn material should be about one-third carbon. So if you had a coal that was 10% ash, we would be taking out 5% carbon with the 10% ash base on coal.

S. VATCHA: I have two questions for Mr. Clark.

What is the status of hot-gas cleanup technology? And how would this affect the market for medium-Btu gas?

PANELIST CLARK: In the case of medium-Btu gas, the effect of a hot-gas cleanup would depend on the end use of the gas. Most uses of gas have difficulty in utilizing thermal content of the gas efficiently. So in most cases, a hot-gas cleanup would give a rather small effect.

If you are going to utilize the gas in a combustion turbine or where you could utilize the heat energy effectively and efficiently, you might have a case for hot-gas cleanup.

S. VATCHA: The second question: Is there a problem with the transportation of high CO gases in a pipeline? Would that require a change in the laws?

PANELIST CLARK: Not to my knowledge. I don't think there is any legal requirement for CO content of gas being transported. Don't forget that for a great many years our household gas had a high concentration of CO. If one would try to use this sort of gas in the home today, I think there would be a serious objection.

TOM R. MARRERO, Visiting Professor, Chemical Engineering Department, Texas A&M University: In nuclear power, they have a problem with regard to solid wastes which are relatively small in amount. What are they going to do with the wastes from the combustion of coal which are apparently mountainous, and also the CO_2? This is to the panel.

CHAIRMAN CONN: Well, I'll try one of them. As you know, one of the problems we have had in Chicago is we can't buy enough salt to put on the streets after the snow. People used to use ash from coal for that, and that is one thing we could do.

Of course, there has been all kinds of talk about the "greenhouse effect" of CO_2. Some of us this winter were wishing we would have a little greenhouse effect to warm up the situation. But, seriously, there is a rising rate of CO_2 concentration in the atmosphere, and exactly what this effect has on the heating and cooling cycles of the world is still under study. I will invite anyone who has some thoughts on that to speak further on it.

T. MARRENRO: I have worked with nuclear power, and I have worked with fossil power, and when I worked with fossil power, there was no way we could find enough uses for the ash from the combustion. In working with nuclear power, one of the problems that they have is that they don't have an acceptable solution for the wastes, and that industry right now is at a standstill.

PANELIST SIEGEL: Let me try to add something to what has already been said about disposal of the ash. The material obviously is quite non-leachable because it has been exposed to some pretty severe conditions in either the furnace or in the gasification process. But leaving open the possibility that some leachable materials might be found in it in the future, that could cause a requirement for needing to fuse the material, to actually slag it, which would make it, of course, much less leachable because of the fused nature. If that ever develops as a requirement, it would give those systems that are slagging systems an extra advantage over those systems that are not slagging and discharge a dry ash, and the systems with the dry ash discharge would have to find some way to add on a final ash slagging step that would make the material almost totally non-leachable. But I think it remains to be seen whether that will be necessary.

PANELIST PASSMAN: We have looked at a number of ashes from different processes, including fluidized beds, and there has been an expectation that there might even be some commercial advantage for highway use. However, the quantities involved are probably beyond what are needed if the use was as widespread as we would like it to be. The handling qualities of the waste are fairly good. We took a look and said, "As a last resort, what could you do?" The plans include putting it back in the mines where the coal came from, and apparently there is adequate space for that. I don't think it is in the same category as nuclear wastes as a handling problem.

CHAIRMAN CONN: So the serious answer to your question then is to fuse it into a material that would not be leachable and put it back in the mines in which there is plenty of room.

T. MARRERO: Then CO_2 remains to be studied.

PANELIST PASSMAN: In answer to the CO_2, it depends on what you are comparing it with. I believe that in most of the processes we have been discussing the CO_2 in the atmosphere is far less than what it would be in a direct burning process. But it certainly isn't zero.

PANELIST CLARK: I think we don't quite understand the CO_2 system yet, the variations of the CO_2 content. This is very closely related to what happens in the oceans and not as closely related to how much carbon we burn or don't burn. So we don't completely understand the entire system. I think it would be premature to try to stop or start or change anything until we understand the system a little better.

CHAIRMAN CONN: But I do think that there have been some curves shown in <u>Science</u> which show that the CO_2 in the atmosphere has been gradually increasing. There is a very important study going on that I have read about, on what the long-range implications would be.

R. BLOOM: I would like to make a comment about the CO in gases being piped around. There may be some local restrictions,

Zeke. I believe Boston has an ordinance controlling the CO content in their pipelines at quite a low level.

But I would like to ask the question of, I believe, Messrs. Clark and Passman. With the attractiveness that Zeke presented on the medium-Btu gas, there seems to be some dichotomy in the attitude of the government. Could you make some comments on this? My question is: Why isn't the DOE a little more active in support of medium-Btu gas programs?

PANELIST PASSMAN: I will answer the part that I can answer. I think that my views and Zeke's parallel one another very closely and we are promoting what we can of medium-Btu gas as rapidly as we can in a commercial sense.

PANELIST CLARK: I think I can give you a little more background. Sometime ago, about three years ago, when ERDA first started, I proposed that we sponsor a large PON, which is a procurement technique, to build a central medium-Btu plant to serve some industrial complex and see if we can split the cost with some industrial partner and get it into the system. And I was told that is so close to commercial realization that DOE shouldn't waste its money on it, that DOE should work on things which require more government help than this system. Well, it's hard to argue with something that you really believe in, so the idea was never allowed to go forward.

So you are right. There is a slight dichotomy here: Should the government sponsor things to the point where industry loses enthusiasm for entrepreneurial effort? There must be some boundary line here where industry has to go and do it itself.

M. WILLINGHAM, Research Analyst, President's Commission on Coal: This is to Mr. Passman. Earlier today, you mentioned that the medium-Btu price would be on the order of magnitude greater because of the oxygen content. Is that strictly the investment cost that you were talking about?

PANELIST PASSMAN: I was talking about investment costs on the order of magnitude of the capital investment over a low-Btu which is on the order of $10 million for typical plant that has been looked at, and $100-$200 million for the medium-Btu plants.

M. WILLINGHAM: And, Mr. Clark, does that square with you pretty well? Do you agree with that?

PANELIST CLARK: You have to be careful here. The $10 million plant will obviously not produce the same number of Btu's as the $200 million plant. The $200 million plant is probably a 60-100 billion Btu's-per-day plant. The $10 million plant is probably something on the order of several million Btu's-per-day plant. I don't remember the exact quantity. But it might involve 300 or 400 tons of coal per day, whereas with the other one we are talking about 5,000 to 8,000 tons of coal per day. So we are entirely in agreement here.

Now, how much it really will cost is a paper study, and one has to know what year it was made, what assumptions were made and what process was assumed. These are not really things that are comparable.

GENERAL CHAIRMAN PELOFSKY: Let me ask a question of Henry Linden. How do we make energy look good to the public?

H. LINDEN: By having a national energy policy which minimizes the cost of energy and takes the crisis out of it. So I would say that playing with energy as a social engineering tool is not the way to do it, and these continuing threats of gasoline rationing, and so on, are not the way to do it. We have technology on the shelf. We could make all the automotive gasoline that we could possibly use at a pump price less than what the Eurpoeans pay for gasoline today. I think we should give the public a chance to do that by removing the institutional and regulatory barriers to do this. Certainly to subsidize oil imports is not the way to do it by entitlements, etc.

I think a good example of good government intervention in energy is the automotive efficiency legislation. A national interest was recognized to increase automotive efficiency, and it was legislated as a standard, and then private industry was allowed to do what they needed to do for it to be done in the most cost-effective way. There are many people who disagree with this approach. But we have no national laboratories on automotive engine development. We have a miracle in front of our eyes. GM, Chrysler, Ford have gone about solving a very difficult problem, and they are doing it without a great deal of government intervention and cost to the consumer. There is a cost to the consumer. but still it is done in the most cost-effective way. To get into an improved energy security situation, you can legislate that 5% of all fuels in interstate commerce in 1990 have to have domestic synthetic sources in them. You can use gasahol, you can use oil shale, you can use coal; you can do whatever you please. Exxon can sell entitlements to everybody else to get the best efficiency of scale. I think it is a totally good process, and it won't cost the public very much. So you have 5% of the gas and oil supply at double the cost of the rest of the supply. It's hardly going to show up in the bill. But the way we are doing it arrives at the most cost inefficiency, technology inefficiency, the most government involvement that we can conceive of, and that's just silly.

I think we can have energy abundance with minimal environmental impact and minimal consumer cost. There are many solutions. But the crisis because of atmosphere, which simply is a means of maximizing government involvement, is not a way to make the public feel good about energy. And to preach to them that energy use is bad, to moralize about energy, is not the way to do it. There are many other things you can moralize about.

GENERAL CHAIRMAN PELOFSKY: I have a problem with your answer, Henry. You say to remove the institutional barriers, and that's fine, except in order to remove the barriers, you need the approval of Congress and that means their constituency must approve that. It's almost the chicken-and-the-egg situation. How do you get what comes first?

H. LINDEN: Well, the public utility energy system certainly has worked very well in the past of having to assume innovation and risk-taking. Certainly the electric utility industry has managed over the years to produce new power sources. If we take the coal gasification issue that Dick Passman talked about, we have a perfect example of how institutional barriers could be removed. We've got a consortium of companies that wants to build a Lurgi plant. That certainly is in the national interest. The only remaining question is: Should the rate-payers of the five pipelines pay for the venture? Should the taxpayers pay for the venture? These are the only two sources of money, right? Or should the entire company and the rate-payers pay for the venture, because it is really not a gas-applied project but a pioneer project?

It seems to me that it doesn't take too much courage to come up with a good solution to that. The stream of 125 million cubic feet per day of $6 or $7 gas and $83 isn't going to break the 25 million customers of those five pipelines, I don't think. You can hardly find it. So that certainly has the capability of being done. But we have all the different state utility commissions and the Federal Power Energy Regulatory Commission and everybody else involved. This is a great exercise in public policy-making, and I think it could be solved by somebody who is reasonably courageous at the top putting his job on the line and saying, "This is how we are going to do it." Many of us put our jobs on the line every day, every week.

G. HL BEYER: I believe you said that the first priority was to be accorded transportation, and I would like to have the panel's reaction to the scenario that if transportation is the most important priority, there is a good chance that fuel oil supplies for heating will dry up in the next ten or twelve years because that fuel oil will be made into gasoline and be outbid by the transportation aspect of the market; and people who are now using fuel oil will experience a rather radical and rapidly escalating price.

PANELIST PASSMAN: First, I would like to clarify my statement. I said that my personal opinion was that transportation fuel will be the first one that would cause a demand for synthetic fuels, because I thought that as a result of current world situations, we might have lines at the pump again and the outcry would be, "With all you $10 billion a year, what have you done for transportation fuels?" I didn't really say it was the first priority nationally. I think that priorities have been set for natural gas which indicates that residential heating is probably going to be the first priority. Before we let people become cold, we will close up a lot of other things.

I think that one of the driving forces for low- and medium-Btu gases by industry, is that they don't want their work interrupted by a stoppage of natural gas supply. They would like to be in control of their energy with their own coal pile and their own generating capability. In these processes, they will have that control.

PANELIST CLARK: You know there is good historical precedent for that: the first time the Office of Synthetic Liquid Fuels was organized. Congress voted for this bill because at the time it came up for a vote, there was a transportation problem in the Washington area and one couldn't get fuel oil delivered. A lot of the Congressmen's homes were cold and they voted overwhelmingly to set up an Office of Synthetic Liquid Fuels. So possibly, as Dick says, that will be what will spark the demand for synthetic fuels.

D. CARLTON: Zeke, I would like to take off from a point you made earlier and end up with a question for Dick.

I am a great believer in medium-Btu gas, as you are. I don't think anybody would disagree with the fact that if we decided this afternoon to build a commercial scale medium-Btu gas plant, we are probably looking at pretty close to ten years before we finally would get some gas out of the other end of that plant. That gets us to the 1990 time frame, and it seems to me that we are finally coming around to the point at which we are recognizing that to look at those kinds of time frames, the kind of price differentials we have, etc., it is tough to make a case for a true freemarket, viable kind of synthetic fuels industry at this stage of the game. I think it is tough for private industry to just launch it with no kind of incentive. And, Dick, it sounds to me as if the government has come to the point where that is recognized.

I would like to throw this question to you. If you take a look at history and the many attempts we have made to get close to a synthetic fuels industry, there has always been something. We started a shale oil industry and then we found a well in East Texas; then we started something else and we found gas somewhere. It looks to me like President Carter is going to Mexico to work out a deal to put a forty-eight line across the Rio Grande. I think that's a swell idea, by the way, and I'm all for that. But my concern is whether we are above to negotiate such a deal. Now we have a "gas black" and now if we have Mexican gas supplies, we are going to find the Department of Energy trailing off in the sunset saying, "We've got other things to worry about." I am afraid that if those pressures come to bear, once again synthetic fuels are going to slip to the background until we run out of Mexican gas, and then we'll start back up again.

PANELIST PASSMAN: Well, all things are possible, but I don't believe so. I think it is true that it is going to take a long time to get a commercial capacity in place to make a significant difference in our supply. A few years ago, the objective was to have a certain capacity of the various fuels by 1985, because that was close enough then that it would have some political meaning. If I can sense the change that's occurred, we are no longer saying that we are going to take over that commercial responsibility.

What we are going to provide is the capability. We are going to provide plants that will be role models that could be replicated. I would rather say they would be a basis for individual companies to project more accurately the cost increments for their own situation. In fluidized bed, for example, it might be a little different in the chemical industry from the petroleum industry and from the steel industry, and it might be wise to have a typical working example in each because various companies have different sizes and operating conditions that would have to be satisfied. Also, estimates in capital cost and all the other things I mentioned this morning could be adequately assessed.

However, the actual commercialization is not to be declared by the U.S. Government, but will occur when the situation is ripe. As I indicated, I don't think anybody knows when that is. With all those economic figures yesterday and today, nobody knows what is going to force those cost curves to cross. I really believe it is going to be an abrupt event which is what usually causes that to happen.

PANELIST CLARK: I would like to comment on that. First of all, if you go by historical precedence, the government will completely abandon any effort if there is an oversupply, a glut, or a new source of fuel. And if it is usable and we have enough, that will be the end. I don't know if we will completely disband the entire Department of Energy. It might not be a bad idea. But at least it will be severely cut back. Now, I am going by historical precendence. This would be my anticipation.

CHAIRMAN CONN: We said several years ago that in order to meet the demand we were going to have to have an Alaskan find of oil every two years. So it is hard to believe that Mexico would supply that much additional oil. It seems to me that chances are very good that we will continue to need these increases, and I don't think that any one source of new oil is going to change the picture that much.

PANELIST SIEGEL: Concerning intermediate Btu gas, I would like to remind the group of something that Zeke Clark mentioned this morning: that there is a major project being studied seriously and deeply by the Carter Oil Company, an Exxon affiliate, to produce intermediate Btu gas using Lurgi technology in the Gulf Coast area. Of course, I cannot predict what the outcome of these studies is going to be. I can't say for sure, therefore, that the project will move forward. But if it does, its timetable is such that we would have a major intermediate Btu gas plant before ten years from now.

PANELIST CLARK: There are other prospects and other projects going forward that will impact within the next four or five years on a movement toward commercialization in the medium-Btu gas picture.

R. WOLK: This question is for Dick Passman.

I am really confused about what it is you are proposing to do. Is it to build plants that will be role models for industrial companies to follow? Will it be that you and your staff will do studies that will give firm cost figures, or will you contract to have other people do studies? Could you just be more explicit?

PANELIST PASSMAN: I am going to clarify it. What we are aiming for is to have industry build plants that will be role models to give a basis and be representative through this first-of-a-kind plant. And these plants are for those technologies that we believe are capable of being commercialized today. So we are saying that the interested companies will need some help to get started, because no one wants to risk his product by using an energy source that he formerly bought as a natural gas which he turns on and off like a faucet. He would first like to see the process operate for a reasonable period of time in circumstances similar to his use. So we are willing to provide some kinds of financial incentives. We would like to know what industry requires those incentives to be.

Now, we have to go back a step and say with the monies that are available, we would like to start the process by providing a larger number of people some money for planning and feasibility studies toward that end. We intend to initiate some plant activities at a later date.

PANELIST HILL: Industrial gas (low- and medium-Btu gas) at the marketplace is directly competitive with high-Btu gas. The industry has been using a lot of natural gas from the interstate and intrastate pipeline systems.

At the time of the 1976-77 severe industrial curtailment of natural gas, as a result of the diminishing supply primarily due to low regulated prices, many people in industry were giving very serious consideration to the fact that they would be continually curtailed in the future from natural gas. So a number of companies entered into contracts for studies and some of them built gasification plants. The Caterpiller plant in York, Pennsylvania, will soon have their plant coming on-line which is the result of their conviction that industry was not going to have a reliable supply of methane for an indefinite period in the future. At that time, all of the intelligent people aware of the situation were predicting that the situation was going to be that way, and that Congress would probably not make a significant move toward the deregulation of natural gas. As you recall, the Natural Gas Policy Act squeaked through by one vote: it was a very close decision. In my opinion, if it had gone the other way, it probably would have been a good many years before that issue would have gotten that close again before the Congress. With the passage of the Act came the present"bubble" of gas.

The Secretary of Energy recently came out urging industry to switch to natural gas whereas, just a few months earlier, the government's position was that there would not be sufficient natural gas. When Congress passed the Natural Gas Policy Act which "deregulated" natural gas, it also passed the Power Plant and Industrial Fuel Use Act which said to industry, "Thou shalt not build any facilities to burn natural gas or oil in the future. They shall be built for coal." Very shortly after that, the Secretary came out and said, "We've got an overdeliverability of natural gas and industry should be using natural gas." Now, if you are an energy user in industry today, I would not envy you if you were trying to make the decision as to whether you ought to be opting for an industrial gasification facility at your plant with this mixture of signals.

I would argue that had Congress not "deregulated" natural gas and had the natural gas curtailment problem continued, many companies today would be ordering industrial gasifier plants as the preferable option to anything else if they can't get natural gas. If they can get natural gas, the price has got to be pretty high before it is economically attractive.

H. LINDEN: I do want to caution you about oversupply of gas. I would say that we should start from the premise that maintaining the current market share of gas of about 25% is a laudible objective in that the old hierarchy of gas supply starting from conventional, both 1948 natural gas through synthetic gas, medium and high, LNG, masking gas, etc., will be cost competitive with an equivalent hierarchy of supplies of liquids and electricity. I hope this is a sound premise that we maintain the 25% market share of gas.

Then let me add up some figures for you. This would mean that in the year 2000 we would have somewhere between 30-35 trillion cubic feet of total gas supply compared with 19-20 today. What will we have in the year 2000? No more than 15 trillion cubic feet of conventional natural gas, including Alaska. That would be very good for the remaining resource base. Certainly we'd have no more than 7 or 8 trillion cubic feet of so-called unconventional gas--that's the highest forecast we have. That 8 and 15 makes 23. Imports: 8 billion cubic feet a day in the year 2000. That's a huge project. LNG, Alaskan gas, Mexican gas. That's 8 billion cubic feet. That would be pretty high. So we have 23 plus 3. We're up to 26. That leaves plenty of room for synthetic gas, high and medium.

If you accept the premise that it is good for the U.S. economy to maintain a 25% primary market share for gas out of 120-130 quads in the year 2000, which seems to be a consensus projection, then we need every bit of gas we can get our hands on, including to get up to 30-31 trillion, 4 trillion cubic feet a year of synthetic gas from coal. That's something like fifty 250 million cubic-feet-a-day, high-Btu gas plants or the equivalent of medium-Btu, each costing $1,500 million or $2 billion total project cost in today's dollars, $100 billion worth by the year 2000 recognizing 1979. The problem is not oversupply, and the problem is not a mixture of gas supplies that are not capable of competing economically with natural and synthetic liquids fuels or electricity. The problem is to maintain the momentum gotten under way by the Natural Gas Policy Act and all the things that Dick Hill and Dick Passman have been talking about. The problem is not oversupply.

T. MORRERO: Figures have been given for the cost of gas of from $3.50 to $7 per million Btu. On what basis has plant availability been established for these figures? And are the current studies considering the materials of construction and proven hardware for plants to last thirty to forty years?

PANELIST CLARK: I can tell you about the state-of-the-art facilities, and I think you can take this with a little bit of caution because SASOL has been operating Lurgi gasifiers for twenty years. They have built up a maintenance and an operating group which is certainly well trained. They report about a 95% on-stream time. What this really means is hard to determine. But for example, in the Great Plains gasification plant, using Lurgi gasifiers, they anticipate a 90% availability, and they have twelve generators and two spares. So they have a pretty good idea of that particular unit. I believe you are using the electrical utility industry's term for availability. Is that correct?

T. MORRERO: Yes.

PANELIST CLARK: I think we can depend on a greater availability of the gas generator than we are even accustomed to with conventional public utility systems, but that only after a period of time needed to develop a proper maintenance and operating group. Obviously, it is not going to happen the first week, the first month, the first year, or the first two years. But I think we can predict availability for at least that system.

As each system is developed and brought into a position where it is ready for commercialization, I hope we will be able to follow Howard Siegel's objective in doing a very careful process development effort which will consider all the features that you have enumerated, and then some.

PANELIST PASSMAN: I would just like to comment also that various speakers indicated that the methods of calculating these things vary from company to company. There is a variation in perception of technical risk and backup, contingency, whether it is a mine-mouth plant or whether coal is transported, whether they need a new pipeline or whether they are using a current pipeline at low capacity with the advantage of the incremental feed into the pipe; and maybe whether there are new experiences to those companies in putting it on the line so that there is a greater uncertainty. There are many factors. Are they using a process in toto that has been proven before? Are they going to need a backup? In other words, what will be the assessment of their reliability? There is a range of prices estimated.

Also, we try to look at it in the same way that FERC looks at it. We look at first-year costs; we look at fifth-year costs; we look at the first five-year average cost; and we look at what people call a levelized cost; and we use it with constant dollars and year-of-expenditure dollars. I am sure that with the various companies looking at it in their own way, there probably aren't two that are done exactly the same. There will be a range.

T. MORRERO: Are the materials of construction in hand today?

PANELIST SIEGEL: I would be glad to try to answer that. The question has been raised often about the process plant design and construction capability in the U.S. Will it have the potential to build as many plants as would be necessary to accomplish a substantial synfuels volume by the year 2000? This has been considered a number of times by a variety of groups. The conclusion has been that the process plant industry and the equipment fabrication industry could grow at a sufficiently rapid rate so that, as an example, by the year 2000, it could put in place 50-100 synthetic fuels plants having an average capacity of 50,000 barrels per day each, which would total 2.5 - 5 million barrels per day of synthetic fuel capacity.

However, this would not be easy. The process plant industry would have to grow at a rate of not just 3% or 4% a year, but more like 8% or 9% per year which they have shown in the past they can do over a sustained period. But the key here is (1) to get started, and (2) to have a clear plan so that the process plant industry knows that the plants will be built. With that atmosphere, they can do the job.

W. R. EPPERLY: I have a question for Messrs. Passman and Hill. Given the lead time that is required to build gas plants, we know that it would not be possible to have a substantial synthetic gas industry until some time in the 1990's. Given that, I would like to submit that the real question is: How much gas is going to be available in the 1990's and beyond in comparison with the demand? There has been a lot of discussion about how much Mexican gas might be available today. I would appreciate your thoughts on how we are going to be able to make projections of the supply and demand well out in the future. And once we have done that, how can that be communicated to the public in a credible way? It seems to me that there is the crux of the problem.

PANELIST HILL: I agree. There is a real problem in trying to make these predictions. In the natural gas area it is particularly difficult because of the situation that has existed for such a long period of time wherein the Federal Power Commission was holding the wellhead price of natural gas at an extremely low and artificial price. The Federal Power Commission, in setting that price, considered only proven reserves of natural gas. In turn, proven reserves are defined as resources that can be produced at prevailing prices. A company would look at a new natural gas development decision on the basis of what gas they could get at prevailing wellhead prices and make a profit.

It has only been in the 1970's that the Commission began to find ways to get out of the court-mandated lock that had been put on previous commissions. The Commission now builds a reasonable incentive into the wellhead price of natural gas. As I have indicated, it was in 1973, only five years ago, that the Commission began to build into its wellhead price the concept of projecting a cost into the future. Previous to that, it was always based on what it actually cost the company to bring in a well. There was no real incentive for developing new supplies.

It was in 1973 that the Commission allowed future projections of cost and the gas price was brought up to about 50¢. This action was tested in the courts and the Commission's approach was sustained. In 1976 the Commission made the dramatic step of doing some real forward-looking projecting and some revolutionary things with the concept of how income tax should be treated and set a wellhead price of $1.50. In the summer of 1977, the Supreme Court denied certiorari. Thus, it has only been since the middle of 1977 that people, producers of natural gas could be comfortable in believing that they could get at least $1.50 for any new gas. Between 1976 and 1977, under the lower court order, they were collecting $1.50 subject to refund of $1. It is not a great incentive when you can collect $1.50, but you have got to put $1 in the bank because you may have to give it back with interest. It was only in 1977 that producers began to see a price which was beginning to resemble a realistic price. Then last year, the price for new gas was raised to $2.

There are all types of predictions on just how much natural gas is truly available in the United States at those prices and how long it will take to develop it. The real problem is trying to get a good handle on it, and I know Henry has been dealing with it in much greater depth than I have been in the last two and a half years. But you are right. How do we get a handle on realistic projections of this gas and then communicate this? That is part of what I was saying before.

The Department of Energy, I'm afraid, has got to take some of the blame for the mixed-up signals that are being sent out to people, particularly people in industry who are buying natural gas and trying to understand what their supply options are going to be and whether they are going to be able to have it.

W. R. EPPERLY: I would like to say that I think all of us here, as well as the people with whom we communicate and work, have the responsibility to stress the lead time that is required. I don't think the lead time is very well understood. At the very least, that will bring about a greater appreciation of the need to try to project what the future supply and demand situation will be on a much more than a one-year or a five-year basis.

In that connection, those of us who are familiar with the technology have a real responsibility not to make establishment of a synthetic fuels industry seem too easy. In our eagerness to say that we can do certain things, I think we have to be realistic about the fact that it is going to take a long time, and be sure that we exercise our responsibility not to make the public think that we are going to overcome this in, say four or five years.

PANELIST HILL: I think that is a very important point, and I agree completely.

PANELIST CLARK: I think I have one thing to add, and that is that I do not believe you will communicate it in a credible manner to the congressional representatives of the general public.

C. LANNING, Project Leader, Department of Energy: I have been a little bugged ever since Mr. Passman commented about liquids for transportation being the top priority in coal liquids. I am happy to hear this, because we at Bartlesville are interested in liquids for transportation-type use. I don't get the impression here that the processes being talked about are for transportation: they are essentially for utility if you want to call them fuel oil. I could go on about this.

Do you see in the activities of the DOE anything to encourage industry to move toward transportation-type liquids from coal? All I know of is a couple of R&D projects to refine liquids.

PANELIST PASSMAN: As I indicated, we are paying attention now to Fischer-Tropsch and coal-to-methanol processes with particular attention toward their transportation potential, and there are other potentials as well. If you are talking about an R&D program and ET, most of the money is toward the direct processes.

PANELIST SIEGEL: I have an additional answer to that. A coal liquefaction process, such as the EDS process that Bob Epperly described to us yesterday, has the capability to produce up to one-half of the total product as a naphtha that could go directly to gasoline, and that certainly fits the description of a transportation fuel. The other half is often called a burner fuel or a utility fuel. But what needs to be appreciated is that when these liquids go into a utility fuel application, it frees up natural petroleum that would otherwise have gone into that application, and that this natural petroleum is then readily convertible through hydrocracking, catcracking, and other normal methods to transportation fuels. So, with EDS liquids, you have half of the product being directly a transportation fuel and the other half being indirectly a transportation fuel because it frees up natural oil that can be processed into transportation fuels.

I think that is an important point to appreciate. And when looked at that way, it says that the total output of a coal liquefaction plant could be considered a transportation fuel. In that situation, that interchangeability can exist for a long, long time, until the point comes when coal liquids become such a major part of the total oil pool that you need to begin converting the coal liquids themselves to lighter products. That will be more difficult to do than converting heavy petroleum materials to lighter products. But, that's a long way off. That is like Step 10 and we haven't taken Step 1 yet. So I don't think it's a real concern at this point.

D. CARLTON: Dick, I can't resist the temptation not to point out to you that while the FPC was fooling around for five years trying to get the price of a buck and a half, the free market intrastate price headed at about two bucks, and in that area.

PANELIST HILL: That's correct. And, of course, this is one of the interesting situations that an agency like the Federal Power Commission has to deal with. During the five years that I was with the Commission, it was an interesting experience because during that time 100% of the Commissioners favored the deregulation of natural gas for the interstate market. There was a turnover with the Commissioners, too. Some had qualifications, but the majority had no qualifications. You had a group of Commissioners who would spend part of their time up on the Hill arguing for Congress to deregulate natural gas; then spending the remainder of their time carrying out their regulatory mandate to control prices under the extremely complicated set of rules and regulations that had evolved since the 1954 Phillips decision. So on the one hand, you had almost a pleading--"Please take this burden from us"--but having to go back and carry out the obligation of the office, which clearly by Supreme Court mandate was to regulate wellhead prices under a strict cost criteria. And it is very difficult to be innovative in a regulatory environment like that. When the Federal Power Commission raised the wellhead price of natural gas from 50¢ to $1.50, there wasn't a single attorney, including the Commissioners, who believed it would survive a Supreme Court test. But it was a matter of "My God, we've got to try something!" And there was nobody more surprised in Washington when the Supreme Court upheld that decision than were the Commissioners and the attorneys for the natural gas industry, the interveners and everybody. It is very difficult to be innovative.

But you are right. In the last few years Commission decision after decision had to deal with the fact that the free intrastate market was paying something like $2 per million Btu, while the Commission was putting together the most horrible patchwork of rules and regulations trying to let a little gas get into the system at deregulated prices. I remember one Commission meeting that I sat through as we were dealing with the loopholes and the loopholes in various emergency procedures the Commission had created to try to relieve the shortfall and the curtailment. I went back to my office and put on the top of my blackboard--and it stayed there for my last two years--a paraphrase, "Oh, what tangled webs we create when we try to regulate."

RECEIVED May 21, 1979.

INDEX

D

E